Recycled Papers: The Essential Guide

Recycled Papers
The Essential Guide

Claudia G. Thompson

American Institute of Graphic Arts / Boston Chapter

The MIT Press
Cambridge, Massachusetts
London, England

Copyright © 1992 Claudia G. Thompson

All rights reserved. No part of this book may be reproduced in any form by any electronic or mechanical means (including photocopying, recording, or information storage and retrieval) without permission in writing from the publisher.

Library of Congress Cataloging-in-Publication Data

Thompson, Claudia G.
   Recycled papers: the essential guide / Claudia G. Thompson.
     p. cm.
   Includes bibliographical references and index.
   ISBN 0-262-20089-9.— ISBN 0-262-70046-8 (pbk.)
    1. Waste paper— United States— Recycling.   I. Title.
TS1120.5.T46   1992   91–46091
676'.282— dc20   CIP

Learning Resources Centre

Centre de ressources pédagogiques

Collège Cambrian College

Sudbury, Ontario

"A slow sort of country!" said the Queen. *"Now, here, you see, it takes all the running you can do, to keep in the same place. If you want to get somewhere else, you must run at least twice as fast as that!"*

— Lewis Carroll, *Through the Looking Glass*

Contents

Contributors	*x*
Acknowledgments	*xii*
About Printing & Writing Papers	*1*

1 The Challenge — 3
The Solid Waste Problem — 4
The Paper Component of the Waste Stream — 7
Paper Consumption — 8
Deforestation and Wood Consumption — 10
Wastepaper Utilization in the Paper Industry — 14
Recycled Printing and Writing Papers: A Marketing Fix? — 16

2 A Brief History of Papermaking — 21
The Invention of Paper — 21
The Spread of Papermaking to Europe and America — 24
The Invention of the Paper Machine — 28
The Development of Wood Pulp — 31
Recycled and Deinked Wastepaper Rediscovered — 33

3 How Recycled Papers Are Made — 37
Wastepaper: A Diverse Resource — 37
Recycled into What Form? — 41
Wastepaper Collection and Utilization — 42
Composition of Recycled Printing and Writing Papers — 44
Deinking Technology and Methods — 48
Deinking Mills — 54
Paper Mills without Deinking Systems — 56
Deinking Sludges — 58
The Office Wastepaper Problem — 59
That Recycled Look: Fiber-Added Papers — 61
Cotton Fiber Papers — 62
A Comparison of Virgin and Recycled Paper Manufacturing — 64

4	**Definitions and Standards**	69
	The EPA Guideline	69
	The Recycling Symbol	74
	State Laws and Regulations	75
	Toward a National Standard	78
	The Great Postconsumer Waste Debate	81
	The National Regulation of Marketing Claims	84
	Private Certifications and Product Evaluations	85
	The Canadian Environmental Choice Program	90
5	**The Characteristics of Recycled Paper**	93
	The Physical Properties of Recycled Fiber	94
	Working with Your Printer	97
	Brightness Standards	103
	Permanence of Recycled Paper	105
6	**The Designer's Legacy: Closing the Loop**	109
	The Paper Specifier's Impact	110
	Creating a Recyclable Product	113
	Soybean Inks	118
7	**Choices and Opportunities**	121
	Comparisons with Other Countries	122
	What You Can Do: A Waste Management Hierarchy	124
	Future Decisions	127
	Appendices	131
1	*Glossary*	132
2	*Pulping and Papermaking Processes*	136
3	*Bibliography & Resources for Further Information*	138
4	*Recycled Papers Available*	146
5	*Designer Impact Analysis Form*	156
	Index	158

Recycle:

To convert waste into a usable form; to reclaim a material from waste.

Paper:

A substance composed of fibres interlaced into a compact web, made (usually in the form of a thin, flexible sheet, most commonly white) from various fibrous materials, as linen and cotton rags, straw, wood, certain grasses, etc., which are macerated into a pulp, dried, and pressed (and subjected to various other processes, as bleaching, colouring, sizing, etc., according to the intended use); it is used in various forms and qualities for writing, printing, or drawing on, for wrapping things in, for covering the interior of walls, and for other purposes.

Essential:

Thorough, entire. Absolutely necessary, indispensably requisite.

— *The Oxford English Dictionary*, 1989

Contributors

The American Institute of Graphic Arts

The American Institute of Graphic Arts is a national nonprofit organization that promotes excellence in graphic design. Founded in 1914, the AIGA advances graphic design through competitions, exhibitions, publications, professional seminars, educational activities, and projects in the public interest.

The Recycled Paper Project

The Recycled Paper Project is a special project begun in 1989 and administered by the Boston Chapter of the AIGA, with the support of the national organization. The project was established to provide accurate and complete information to designers, paper specifiers, and the public, and to enable all individuals involved in the production of printed materials to make environmentally responsible decisions. This book, which is the product of that project, was made possible by generous contributions of time, money, talent, advice, and assistance from a tremendous number of individuals and organizations throughout the United States. We gratefully thank all of them. Their extensive support enabled us to take a comprehensive approach to the content of the book, and to print and distribute copies as an educational resource to the entire AIGA membership.

Recycled Paper Project Team

Claudia Thompson
Founder & Director

Susan Larocque
Project Assistant

Contributing Assistants
Lory Christoforo
Christa Garubo
Jean Hammond
Lenore Lanier
Pam Simonds

Wendy Gerlach, *Ropes & Gray*
John Manopoli, *Ropes & Gray*
Matthew Wick, *Coopers & Lybrand*

Review Committee

Geoffry Fried
Vice President, AIGA/Boston

Caroline Hightower
Director, AIGA

Allen Payne
Director of Programs, AIGA

Anthony Russell
President, AIGA

Douglass Scott
*Senior Designer,
WGBH Educational Foundation*

Judith A. Usherson
Center for Earth Resource Management Applications

**AIGA/Boston
Board of Directors 1990–92**

Terry Swack, *President*
Clifford Stoltze, *Vice President*
Geoffry Fried, *Vice President*
Jean Hammond, *Treasurer*
Jim Efstathiou, *Secretary*

**AIGA/Boston
Board of Directors 1988–90**

Judith Richland, *President*
Geoffry Fried, *Vice President*
Kim Halliday, *Treasurer*
Jennie R. Bush, *Secretary*

Project Advisory Board

Jerry French
*Executive Vice President,
French Paper Company*

Steven Heller
Senior Art Director, The New York Times

Tibor Kalman
President, M & Co.

Susan Kinsella
Californians Against Waste Foundation

Jobe B. Morrison
President, Miami Paper Corporation

Thomas Ockerse
*Head, Graphic Design Department
Rhode Island School of Design*

Stephanie Pollack
Attorney, Conservation Law Foundation

Jerry Powell
Editor, Resource Recycling

Douglass Scott
*Senior Designer,
WGBH Educational Foundation*

Judith A. Usherson
Center for Earth Resource Management Applications

Project Funding

Major funding for this project was provided by generous grants from:

The National Endowment for the Arts
The Joyce Mertz-Gilmore Foundation

Partial funding for the printing and distribution of copies to AIGA members was provided by:

The American Institute of Graphic Arts and the following AIGA Chapters: Boston, Chicago, Cleveland, Honolulu, Indianapolis, Kansas City, Knoxville, Los Angeles, Miami, Minnesota, Nebraska, New York, Philadelphia, Portland, Rochester, San Diego, Seattle, Texas

Additional funding to support project research and book production was provided by the contributors listed below.

Patrons ($5,000–9,999)
Aldus Corporation
Environmental Protection Agency: *(Assistance Agreement X1001499-01-2)*

Sponsors ($2,000–4,999)
Beloit Corporation
Boston Globe Foundation
Center for Creative Imaging
Kodak Electronic Printing Systems
Lotus Development Corporation
R.R. Donnelley & Sons
3M Corporation

Associates ($1,000–1,999)
Margaret Boothe
Braceland Brothers Printing
Communication Arts
Houghton Mifflin Company
Little, Brown and Company
Jean R. Thompson

Friends ($500–999)
Clement Mok Designs
Clifford Selbert Design
The Duffy Design Group
Hawthorne/Wolfe
HOW Magazine
M & Co.
The Mednick Group
The Nimrod Press
Pantone
Vignelli Associates

Subscribers ($250–499)
Allemann Almquist & Jones
Charrette Corporation
Cook & Shanosky Associates
Lloydesign Associates
Mueller & Wister
Felice Regan/The Graphic Workshop
Sherman Design
Sibley/Peteet Design
Step-By-Step Graphics
Sumida Design
Tim Girvin Design

Colleagues ($100–249)
Ecoprint
April Greiman
Jean Hammond

In-Kind Donors
Adobe Systems
Anderson Fraser Publishing
Art Related Technology/Stats For U
Century Color Company
Coopers & Lybrand
Cross Pointe Paper Corporation
Daniels Printing Company
French Paper Company
Gifts In Kind America
Harvard Pinnacle Group
Landslides/Alex S. MacLean
Judy Love
Miller Freeman Publications
Mohawk Paper Mills
Monotype Composition
PageWorks
Paper Recycler
Printing Complaints Incorporated
Recycled Paper Company
Recycled Paper News
Ropes & Gray
Stanley Rowin Photography
Simpson Paper Company
The Stinehour Press
TAPPI Press
Typotech Reproduction
Weymouth Design
Zona Photographic Laboratories

Project donors did not assume any editorial control over the writing or production of this work, and are not responsible for its content. The views expressed do not necessarily reflect the views or policies of the organizations and individuals that provided funding. Nor does the mention of trade names or commercial products constitute any endorsement or recommendation for use.

Acknowledgments

I wish first to thank my husband, Roger Boothe, though it is difficult to do so adequately. His support has been instrumental to my completion of this work. If all of humankind shared his generosity and integrity, we would be in good shape. My assistant, Susan Larocque, also deserves a great deal of credit. Her dedication and good humor pulled us through many long, difficult days of work; her design talents did much to shape this book. I am also indebted to Larry Cohen, my editor at MIT Press, who did much to improve the prose; it was a pleasure to collaborate with him. Paul Hoffmann and the staff at The Stinehour Press are responsible for the fine quality of the printing.

Several members of our Advisory Board devoted significant time and energy to this project— Susan Kinsella, Jobe Morrison, Douglass Scott, and Judy Usherson. At the American Institute of Graphic Arts, Geoffry Fried provided valuable comments and ideas. Joe Duffy, Caroline Hightower, and Anthony Russell all thoughtfully reviewed the first draft of the manuscript.

Many cultural institutions across the country made their resources available. Robert Rainwater of the New York Public Library and Helena Wright of the Smithsonian Institution were particularly helpful; they reviewed materials and graciously opened the doors to their collections. Others in varying capacities gave generously of their time— Richard Braddock of the EPA, Jeff Coyne and the staff at Earthworm, Ted Vansant of the Recycled Paper Company, Ellen McCrady of the *Alkaline Paper Advocate,* Chandru Shahani at the Library of Congress, Debra Riccio and the staff at Daniels Printing, Mitra Khazai, Tom Soder, and Nancy VandenBerg.

The cooperation of many individuals in the paper industry and related fields, who patiently shared their expertise and knowledge, was essential to this work. Charles Klass was particularly generous. It is not possible to list everyone who offered help, but I would like to thank David Assmann, Ed Atwell, Wally Bergstrom, Chuck Birk, Dave DeYoung, John Erickson, Tom Garbutt, Judy Gowdy, Andy Harrison, Paul Kreeger, Tom Meersman, Charles Moyer, Jr., Daniel Rogers, Joyce Pekala, Tony Rubio, John Schulte, Tom Sullivan, Glen Tracy, and Edward Zajak.

Alice Ledogar of the Conservation Law Foundation was especially helpful to our fundraising effort; Rachel Pohl and Don Falk also assisted with these matters. Our proposal and budget were ambitious, so I thank Suzanne Watkin and The Boston Globe Foundation for being the first to take the leap and express the confidence of a financial contribution. Our very generous funding from the Joyce Mertz-Gilmore Foundation allowed us to develop the contents in a thorough manner; it was enjoyable to work with both Robert Crane and Penny Willgerodt. The National Endowment for the Arts was most generous; we could not have accomplished the AIGA distribution without their support. Thanks also to Ron Jennings and EPA Region I, to Deirdre Devlin and Aldus Corporation, to Jean Thompson and Margaret Boothe, and to all the other donors who made this possible.

There are others who deserve recognition. Ted Smith and Lynn Anderson paved the way with *Paper Chase II* and always offered encouragement and advice. Sharon Poggenpohl, Eva Anderson, and Jeanne Lee all contributed insight. Judy Love generously drew several illustrations. Tessa Huxley and Andy Reicher provided thoughtful comments along with shelter in the Big Apple, which made the traveling much easier. And Doreen Arcus helped me see the light when, at the last possible moment, we had to commit to a title. I hope this book will make a constructive contribution to our future, and I am grateful to everyone who helped the work come to fruition.

About Printing & Writing Papers

The products of the paper industry are often classified by their end uses into five principal categories: paperboard, printing and writing papers, newsprint, tissue, and packaging and converting papers. Paperboard, which includes the heaviest weight paper products— principally boxboard and the containerboard used to make corrugated packaging— makes up the largest class of paper products consumed in the United States. Printing and writing papers are the second-largest category of paper products. The majority of these papers are used for offset printing; however, this diverse group also includes office papers used for photocopying or other duplication processes, stationery and tablet papers, forms bond papers, envelope converting papers, and papers intended for a few other uses.

Newsprint is the third-largest category of papers consumed— it is given a separate classification because it represents a product that is consistently manufactured for a single printing purpose. Fourth in consumption ranking are tissue products— principally toilet and facial tissues, toweling, napkins, and other sanitary products. The final category of products is packaging and industrial converting papers, including bleached and unbleached papers used for bags, sacks, and other packaging, as well as some specialty papers.

The focus of this book is on printing and writing papers. Unless otherwise noted, the detailed discussion of manufacturing processes and related issues applies primarily to these products. Additional information is provided about the paper industry as a whole. This book does not address the subject of making recycled newsprint, as developments in this area are quite different than they are for printing and writing papers. Some of the information contained here will, however, apply to packaging papers (especially bleached ones) since production processes for these papers are similar to those used for the manufacture of printing and writing papers.

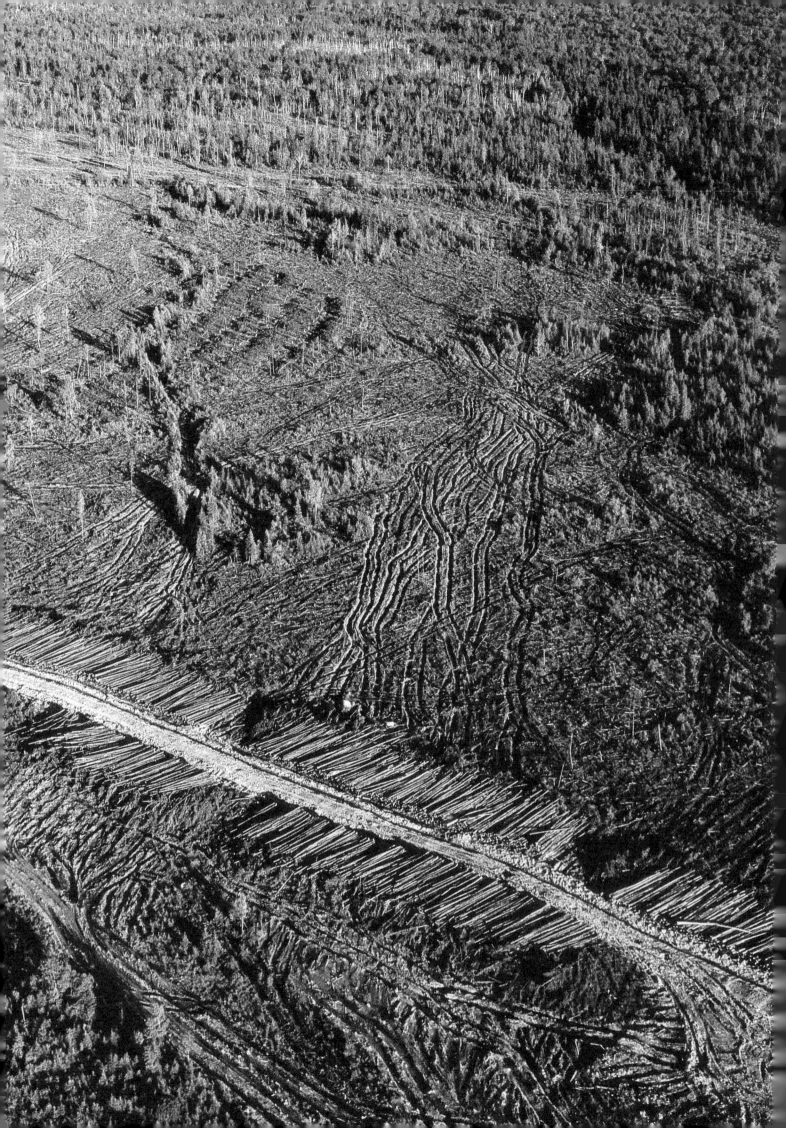

1 The Challenge

As we approach the end of the twentieth century, we face a mountain of environmental problems wrought by accelerating technological development and increased consumption patterns. We humans know, both intellectually and intuitively, that we are pushing the limits of what our planet can stand. The larger issues before us— global warming, atmospheric pollution, rainforest destruction, among others— have emphasized the importance of all environmental issues and the urgent need for action. The challenge we face as we step into the twenty-first century is to bring our technologically intense civilization into balance with the natural cycles that give life to this beautiful and delicate planet. It is not an easy task.

Of the many difficult problems before us, that of solid waste is one of the easiest to solve. Already, recycling is a popular issue. The changes required of each of us, individually, to make it work are relatively small. Unfortunately, there are also obstacles to this much-needed transition. Despite the current hype and promotional claims, the use of wastepaper in the printing and writing segment of the U.S. paper industry is still very limited. Mill trimming and converting wastes supply a substantial percentage of the "recycled content" for many of these papers. Because we lack meaningful national standards or regulations to govern marketing claims, paper specifiers need to become experts on this issue in order to make purchasing decisions that support expanded markets for recycled wastepaper. Given that printing and writing papers are the largest category of paper products consumed in the country except paperboard, and that their consumption continues to grow and is now almost double that of newsprint, the need for change is real. This book was written to clarify the issues, document the facts, and help all of us become informed buyers— in the hope that we will achieve a substantial increase in recycling rates for these papers.

Logging for pulpwood in Maine leaves skidder tracks over the landscape where trees are clear-cut. Timber is stacked in rows for transport to the mills. (Photo © Landslides/Alex MacLean.)

Paper recycling can play a vital role in the management of many of our environmental problems. In addition to conserving resources that have already been extracted and helping to reduce our waste disposal problems, there will be other benefits. In paper manufacturing, as in other industries, using recycled materials to create new products consumes less energy and generates a great deal less overall pollution than does manufacturing using virgin materials. But it is the escalating cost of solid waste disposal, fueled in turn by the environmental consequences associated with our disposal methods and the increasing scarcity of disposal capacity, that has driven this issue to the forefront of our national attention. Tipping fees— the cost per ton for waste haulers to unload their cargo at landfills, incinerators, or transfer stations— have risen dramatically in the past decade, increasing from $25–30 per ton to over $130 per ton for some communities in the northeast. In Philadelphia, for example, the cost of waste disposal rose above the city's annual costs for fire protection, reaching $40 million in the 1988 fiscal year.[1] We are running out of places to put our garbage at almost any price.

The Solid Waste Problem

Americans hold the dubious distinction of creating more municipal solid waste— waste collected from residential, commercial, institutional, and a limited number of industrial sources— than any other nation on earth, whether the quantities are measured in total amounts or on a per capita basis.[2] In 1988 the nation generated a total of 180 million tons, or an

[1]
Louis Blumberg and Robert Gottlieb, *War on Waste* (Washington, DC: Island Press, 1989), p. 125. This book does an excellent job of reviewing the evolution and history of municipal waste disposal in this country as well as our present options.

[2]
Municipal solid waste does not include municipal sludges, incinerator ash, and many industrial process wastes, even though these may also end up being disposed of in municipal landfills or incinerators.

Figure 1.1
Generation of MSW in the United States

Our generation of wastes has grown rapidly over the past three decades, more than doubling since 1960. Paper has steadily increased its share of the total, rising from 34% of all wastes, to the current 40% over the same period. By the year 2000, paper and paperboard products are projected to comprise 45% of all MSW we generate, and by the year 2010, 48%. (Data from Franklin Associates, Ltd.)

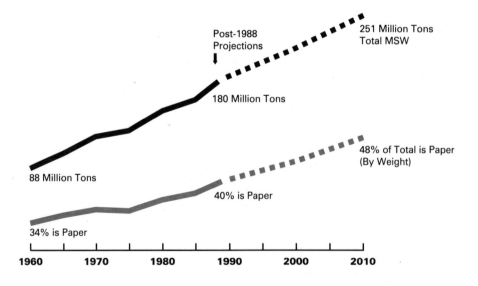

Recycled Papers: The Essential Guide

average of 4.0 pounds per person each day. This more than doubled the 88 million tons generated in 1960, which represented a more modest 2.7 pounds per person per day. Without source reduction, the Environmental Protection Agency (EPA) projects that we will produce 216 million tons of municipal solid waste (MSW) annually by the year 2000, an impressive 4.4 pounds per person daily.[3]

In 1988 only 13% of all wastes generated in the United States were collected for recycling, while approximately 14% were combusted and 73% were landfilled.[4] The environmental problems associated with both the landfilling and the incineration of wastes have become increasingly apparent in the past few decades. Our landfills have become major sources of contaminated groundwater. Leachates are frequently generated containing hazardous organic chemicals and heavy metals such as lead, mercury, and cadmium.[5] The EPA has found potential or existing groundwater contamination problems at the majority of all the landfills it has studied. Contaminants in the leachates of some landfills have reached such toxic levels that 290 sites on the Superfund National Priority List for hazardous waste cleanup are MSW landfills— this represents 23% of all SNPL locations.[6] There is a growing recognition that while better construction methods, such as the use of clay caps and liners, are important, they are no guarantee of safety. Because of these environmental risks and because they have reached capacity, many landfills have been closed. Only about 6,000 of the 20,000 landfills operating in 1978 remained open in 1990— a closure rate of 70% over twelve years. By 1993 the number of operating landfills is expected to be reduced to 3,500.[7]

With pressure growing to find new disposal options for our garbage, the 1980s saw a resurgence in the construction of incinerators. In the coming decade many more communities will face the difficult decision of whether or not to build new facilities. Most of the newer incinerators are being promoted as "waste-to-energy" plants or, even more confusingly, "resource recovery" plants. Energy, usually in the form of electricity, is generated during the burning of wastes. Unsorted municipal solid waste is not particularly efficient fuel, however, and little or no recovery of wastes takes place at these facilities. In addition, the mass burning of the diverse materials in MSW, including household hazardous wastes along with

---

[3] Franklin Associates, Ltd., *Characterization of Municipal Solid Waste in the United States: 1990 Update* (Washington, DC: U.S. EPA Office of Solid Waste, June 1990), pp. ES-2–3, ES-9.

[4] Ibid., p. 74.

[5] Blumberg and Gottlieb, *War on Waste*, pp. 3–21; Richard A. Denison and John Ruston, eds., *Recycling and Incineration: Evaluating the Choices* (Washington, DC: Island Press, 1990), pp. 5–6.

[6] Memorandum and follow-up interview with Sally Mansur, U.S. Environmental Protection Agency, Region I, July and September 1991.

[7] National Solid Wastes Management Association, *Landfill Capacity in the Year 2000* (Washington, DC, 1989); Memorandum from Sally Mansur, U.S. EPA, July 1991.

Figure 1.2
Composition of Municipal Solid Waste Generated: By Weight

Paper is the largest single category of wastes generated in the United States—in 1988 it was 40% by either weight or volume. By weight, yard wastes come in second, followed by metals, plastics, and glass. If measured by volume instead, paper remains the same. Plastics would then be the second-largest category. (Data from Franklin Associates, Ltd.)

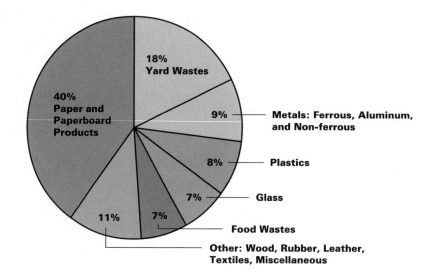

8

For a more complete discussion of the problems associated with incineration see *War on Waste* and *Recycling and Incineration,* already cited. See also Stephen Lester and Brian Lipsett, *Solid Waste Incineration: The Rush to Burn* (Arlington, VA: Citizens Clearinghouse for Hazardous Waste, 1988).

9

Memorandum from Sally Mansur, U.S. EPA Region I, July 1991.

paper, plastics, metals, and other trash, can itself cause serious environmental problems. The remaining incinerator ash, which often has high enough concentrations of heavy metals and other toxic compounds to be classified as hazardous waste, must be disposed of in a landfill. Significant amounts of air pollutants such as dioxins, acid gases, heavy metals, and CO_2 greenhouse gases are emitted during combustion. Fly ash, which is supposed to be captured by air pollution control equipment, can in fact sometimes escape and settle as particulate matter over surrounding communities.[8]

Problems such as these have caused many recently planned incinerator projects to be delayed or canceled. But an even bigger concern to communities is the "put or pay contract." Because incinerators are so costly to build ($50–400 million is a typical estimate) and to operate, the large corporations that manage them have required municipalities to commit to long-term contracts, often 20 years, for the delivery of minimum quantities of wastes. Should a community implement a successful recycling program, resulting in a reduction of wastes below contracted levels, it would still be required to pay the tipping fees for its contracted tonnage. Some contracts have even limited community recycling rates to fixed levels such as 15%.[9] Thus the building of new incinerators can come into direct competition with the development of recycling programs, and the opportunity to develop markets for waste materials is then lost. Conversely, because paper is the largest single component of the solid waste stream, a substantial growth in paper recycling rates can potentially reduce the extent of our need to build new incinerators over the coming decades.

The Paper Component of the Waste Stream

The portion of the waste stream that is paper has been increasing steadily, and paper and paper products make up the largest single category of wastes in the United States. In 1988, whether measured by weight or volume, 40% of all MSW generated in this country was paper— including newspapers, books, magazines, office papers, other commercially printed papers, corrugated containers, paperboard, and paper packaging. After recovery of various wastes for reuse or recycling, paper products still make up a much larger share of the total remaining for disposal than does any other material— glass, plastics, or metals. By 2010 paper and paperboard are projected to comprise 48% of all MSW.[10]

It is a common but mistaken perception that newspapers are one of the largest categories of paper wastes. It is the uniform and readily identifiable nature of newsprint that leads to this false assumption. *A prominent paper industry consulting firm reports that printing and writing papers are the largest single component of wastepaper in MSW and that the discards of printing and writing papers (after recovery of any materials for recycling) roughly equal the combined discards of old newsprint and old corrugated containers.* This firm projects that, if present trends continue, by 1995 the discards of printing and writing papers will be about triple those of newsprint and almost one-and-a-half times the combined discards of old newsprint and old corrugated containers.[11]

[10]
Franklin Associates, Ltd., *Characterization of Municipal Solid Waste in the United States: 1990 Update*, pp. 9–12, 58, 81–91. While statistics for the percentage of solid waste that is paper are relatively constant by either weight or volume, other materials, most notably plastic and aluminum, will appear to constitute a very different percentage of the total depending on whether weight or volume measurements are used. For example, in 1988 plastic wastes generated in the United States represented roughly 8% of the MSW stream by weight, but probably over 17% by volume.

[11]
Fred D. Iannazzi and Richard Strauss, Andover International Associates, "Problems and Opportunities in the Utilization of Recycled Printing & Writing Paper," paper presented at *New England TAPPI/Connecticut Valley Pima Technical Seminar, Holyoke, MA* (March 1991).

Figure 1.3
Printing and Writing Papers Discarded in MSW

In 1988 the total discards of printing and writing papers into the waste stream roughly equaled the combined discards of old newsprint (ONP) and old corrugated containers (OCC). Recycling rates are growing for newsprint and containerboard. Meanwhile, relatively few printing and writing papers are recycled and their consumption continues to increase. Thus, by 1995, the total discards of printing and writing papers will be about triple the discards of either ONP or OCC. (Data from Andover International Associates.)

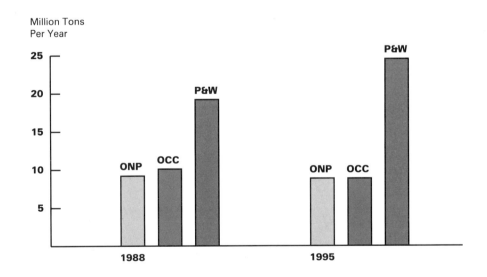

The Ephemeral Products of Graphic Design

A detailed examination of the 1988 data outlining the composition of wastes generated in the United States shows that the products routinely created by graphic designers working with their clients— such as magazines, catalogs, annual reports, books, letterheads, posters, direct mail, other promotional materials, and paper packaging— make up a very significant 13% of the total waste stream. While these designed paper products may all look somewhat different because they are printed on various grades of papers, in a variety of shapes and forms, their combined weight was an annual 23.5 million tons. In contrast, newspapers, which constitute less than one-fifth of all paper products consumed, represent just over 7% of the municipal solid waste total, or only 13 million tons.[12]

Packaging is sometimes reported as comprising the largest single category of wastes generated, and packaging materials do indeed make up 31.6% of the total MSW stream generated, before recycling takes place. But it is important to note that this packaging statistic includes many different materials— all glass bottles and jars, steel and aluminum cans, plastic containers— and that over one-half of what is included in this category consists of corrugated boxes and other paper packaging.

Paper Consumption

A look at consumption patterns makes it obvious why paper is the biggest ingredient of the garbage pile. Americans consume more paper in total amounts and on a per capita basis than the residents of any other country in the world. Excluding paperboard products, the total quantity of paper consumed in the United States in 1990 was 50 million tons, including 25 million tons of printing and writing papers. Newsprint consumption was next at 14 million tons. Tissue products weighed in at 6 million tons, and packaging papers were just behind at 5 million tons.[13]

There has been a historical pattern of continuously increasing paper consumption, evident since paper was first invented, but especially since the development of methods for pulping wood. Despite the recent advent of new communications technologies, this trend has not abated. From 1960 to 1990 the combined U.S. consumption of printing and writing, newsprint, and tissue papers more than doubled. But it was the increase in use of printing and writing papers that was most significant. During those thirty years, annual consumption in this category tripled, rising from

---

12
Franklin Associates, Ltd., *Characterization of Municipal Solid Waste in the United States: 1990 Update,* pp. 24–50.

13
American Paper Institute, *Paper, Paperboard & Wood Pulp: Monthly Statistical Summary,* Vol. 69, No. 1 (January 1991), pp. 3, 9. U.S. consumption data are calculated by adding imports and subtracting exports from U.S. production statistics. Included in the category of printing and writing papers are many types of papers specified for printing jobs (especially coated and uncoated offset papers), which make up the bulk of the tonnage or roughly three-quarters of the total. The remaining one-quarter is composed of a diverse group of papers used more directly for office and other uses, such as tablets and stationery, computer paper, forms bond, converting papers for envelope manufacturing, and several other grades.

Figure 1.4
Graphic Design Products in the Waste Stream

A detailed look at the 1988 data for wastes generated in the United States shows that the categories of products produced by graphic designers make up, by weight, at least 13% of the entire MSW stream. In this chart, these categories are shown in blue. (Data from Franklin Associates, Ltd., noting that tenths of percents may not add perfectly to a 100% total due to rounding.)

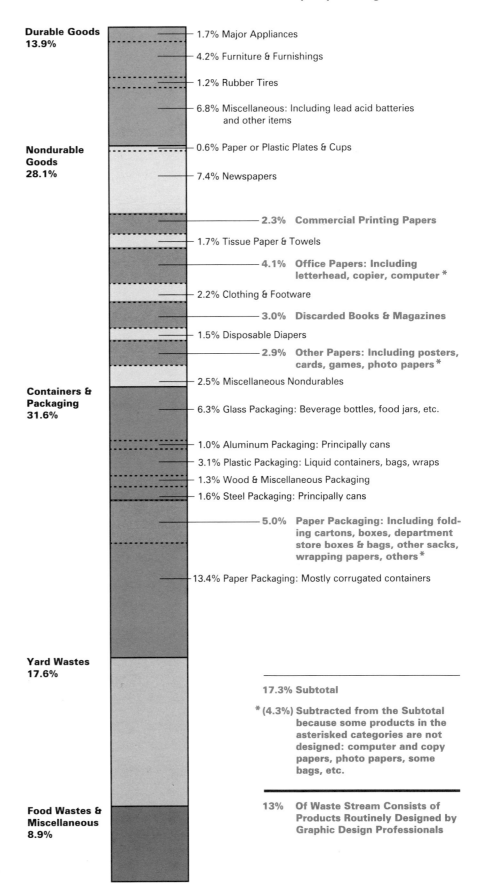

Categories of Wastes That Include Primarily Graphic Design Products

Durable Goods 13.9%
- 1.7% Major Appliances
- 4.2% Furniture & Furnishings
- 1.2% Rubber Tires
- 6.8% Miscellaneous: Including lead acid batteries and other items

Nondurable Goods 28.1%
- 0.6% Paper or Plastic Plates & Cups
- 7.4% Newspapers
- **2.3% Commercial Printing Papers**
- 1.7% Tissue Paper & Towels
- **4.1% Office Papers: Including letterhead, copier, computer ***
- 2.2% Clothing & Footware
- **3.0% Discarded Books & Magazines**
- 1.5% Disposable Diapers
- **2.9% Other Papers: Including posters, cards, games, photo papers ***
- 2.5% Miscellaneous Nondurables

Containers & Packaging 31.6%
- 6.3% Glass Packaging: Beverage bottles, food jars, etc.
- 1.0% Aluminum Packaging: Principally cans
- 3.1% Plastic Packaging: Liquid containers, bags, wraps
- 1.3% Wood & Miscellaneous Packaging
- 1.6% Steel Packaging: Principally cans
- **5.0% Paper Packaging: Including folding cartons, boxes, department store boxes & bags, other sacks, wrapping papers, others ***
- 13.4% Paper Packaging: Mostly corrugated containers

Yard Wastes 17.6%

Food Wastes & Miscellaneous 8.9%

17.3% Subtotal

* (4.3%) Subtracted from the Subtotal because some products in the asterisked categories are not designed: computer and copy papers, photo papers, some bags, etc.

13% Of Waste Stream Consists of Products Routinely Designed by Graphic Design Professionals

The Challenge

New communications technologies, promoted on the basis of ever increasing speed of output, have contributed to great increases in paper consumption. The IBM 1460 computer, first sold in 1963, was noted for printing 1100 lines per minute. (Photo courtesy of the Computer Museum, Boston.)

14
American Paper Institute, *1990 Statistics of Paper, Paperboard, and Wood Pulp* (New York: American Paper Institute, 1990), pp. 26–29, with additional data for 1960 provided by API staff.

15
Franklin Associates, Ltd., *Characterization of Municipal Solid Waste in the United States: 1990 Update*, p. 65.

16
William U. Chandler, *Materials Recycling: The Virtue of Necessity*, Worldwatch Paper 56 (Washington DC: Worldwatch Institute, October 1983), p. 9, citing U.S. Congress Office of Technology Assessment, *Wood Use* (Washington, DC: U.S. Government Printing Office, August 1983).

17
U.S. Department of Agriculture Forest Service, *An Analysis of the Timber Situation in the United States: 1989–2040* (Fort Collins, CO: Rocky Mountain Forest Range Experiment Station, December 1990), p. 42.

8 million to 25 million tons. In the period of only six years from 1982 to 1988, the growth was the most dramatic, with consumption rising from 16 million to almost 25 million tons.[14] The rapidly expanding use of computers, copiers, and facsimile machines has played a significant role in these increases. Futuristic predictions of "the paperless office," which glorified the wonder of these new technologies, have proven naive. They failed to account for the propensity of people to reproduce many more documents at more frequent intervals, given the ease of use and ready accessibility of these machines. During the past thirty years the growth of magazine publishing, direct mail, and the use of printed promotional materials has also contributed to these large increases in paper consumption. This trend shows no sign of abating, as post-1990 projections show that "books and magazines, office papers, and commercial printing are projected to increase their share of total generation more rapidly than other products."[15] Clearly, those of us who are responsible for purchasing and using these papers can play a significant role in solving our solid waste crisis.

Deforestation and Wood Consumption

Paper production consumes a substantial portion of the national and global wood harvest. In 1983 it was estimated that the production of paper products required roughly 35% of the commercial wood harvest worldwide, and this share was projected to grow to 50% by the year 2000.[16] In the United States 27% of the 1986 timber harvest was destined for domestic pulpwood production.[17] Moreover, the United States is one of the world's leading exporters of wood products, and some of our exported timber may end up in paper as well.

In the course of this century our forest practices have evolved from the harvesting of stands of native timber from vast natural forests containing many diverse species, to an increasing dependence on wood harvest from lands that are growing fewer species of trees in a vastly altered environment. There are huge tree farming operations based on "monocultures" for the rapid growth of single species of trees. There are reforested areas that have been seeded with a single species of tree after clear-cutting, and there are second-growth forests that have sprouted after extensive cutting in earlier industrial times. To a varying degree, large quantities of herbicides, pesticides, and chemical fertilizers are applied to these woodlands for their management. The increased emphasis on single-species forests leads to a lack of species diversity and ecological stability overall; it reduces the natural habitat for wildlife, while making the trees themselves increasingly vulnerable to a variety of diseases and insect infestations. Clear-cutting is still a common practice in present-day wood harvesting in all regions of the country, whether the source of the wood supply is native forests or managed timberlands.

The Pacific Northwest supplies one-half of the lumber used in the United States, and about 17% of the pulpwood used in paper production. Much of this is in the form of wood chips, some of which are generated as by-products from lumbering operations. Public lands in the United States, principally those under the jurisdiction of the U.S. Forest Service, currently provide 22% of the country's wood harvest.[18] While many of our national forests are being cut, it is the ones in Washington and Oregon that are being most devastated, and that supply the majority of the timber taken from the public trust. These national forests are being cut at a rate of 240 acres per day, and much of the land area is being clear-cut. Yet it is these national forests that contain almost all of the country's remaining native forests. The clear-cutting is typically followed by slash burning and aerial spraying of herbicides to kill off the residual seeds and roots of the original diverse vegetation. Subsequent replanting is done with a single species of tree destined for future harvest in a relatively short time frame. Meanwhile, the high surface run-off levels during rainfalls cause significant erosion and loss of nutrient-rich soil.[19]

18
Pulp & Paper North American Factbook: 1990 (San Francisco: Miller Freeman Publications, 1990), pp. 57–63.

19
Jeffrey L. Chapman, "Forests Under Siege," *USA Today* (March 1991), pp. 17–20.

National forest lands continue to be clear-cut for their timber harvest; this mountainside in northwestern Montana was denuded in 1990. (Photo © Landslides/Alex MacLean.)

Figure 1.5
**Trends in United States
Paper Consumption**

Consumption of paper and paper products continues to rise dramatically; it reached a total of 87 million tons in 1990. Over the last ninety years, the annual U.S. consumption has grown twelve times on a per capita basis, from 59 pounds in 1900 to 695 pounds in 1990. The pie chart illustrates the relative percentage of each type of paper product consumed today. Printing and writing paper consumption is approximately double that of newsprint, representing almost 30% of all paper products used. (Data from American Paper Institute.)

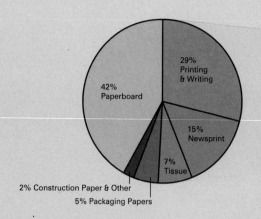

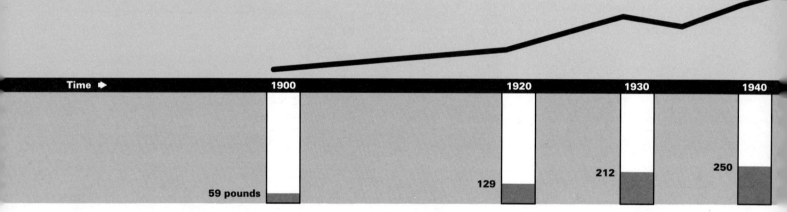

1880
Cotton and linen rags have been the primary fiber used for papermaking. Groundwood pulp, first used in the 1860s, progressively becomes more important.

1882
Sulfite pulp first made in the U.S.

1899
Toilet paper in roll form is now widely used.

The first extensive use of electric power in paper mills.

1903
Corrugated containers are used commercially.

Ladies Home Journal is the first magazine to reach a circulation of over one million copies per year.

1906
Commercial production of handmade paper ceases, until its brief revival starting in 1928 by Dard Hunter, and a later renaissance in the 1970s.

1907
An unexpectedly heavy run of toilet paper is converted to become the first paper towels instead.

1908
Henry Ford introduces the Model T.

1909
Kraft pulp first made in the U.S.

1916
A major shortage of pulp leads Secretary of Commerce Redfield to ask the public to save old paper and rags.

1924
Kleenex facial tissue is introduced.

1928
Teleprinters and teletypewriters come into use.

1930
The supermarket is increasingly important due to the growing use of automobiles and a shift from bulk to packaged products.

1932
In the depths of the Depression, one-half of newsprint mills go bankrupt; 13.7 million people unemployed in the U.S.

1936
Use of paper packaging for milk containers is now widespread.

Phototypesetting becomes available in the U.S.

Life magazine is introduced, selling over 500,000 copies in its first few weeks—more than any other magazine had in its first year.

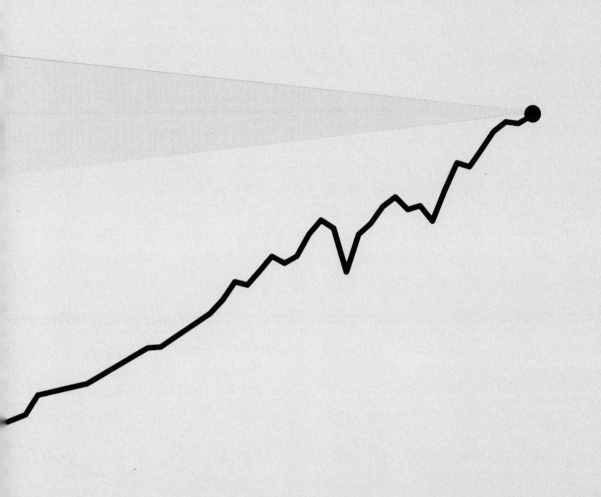

Total Annual Consumption Of Paper Products

90 million tons

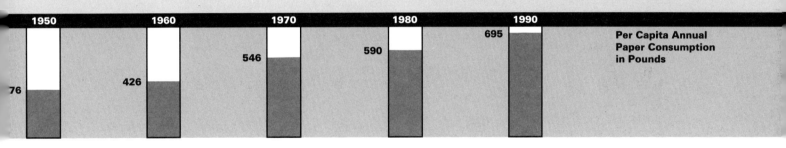

Per Capita Annual Paper Consumption in Pounds

1938
The 40-hour work week is established; one hundred years earlier it had been 78 hours.

1948
Market acceptance of frozen orange concentrate leads to the expansion of the frozen foods industry, with associated increases in packaging.

1951
The first computer is made for commercial purposes; color television first introduced in the U.S.

1953
Watson and Crick discover the double helix structure of DNA; International Paper begins a genetic tree research program.

1954
The wastepaper utilization rate for printing and writing papers is over 20%. It will decline to 6% in 1990.

1959
The first photocopier, the Xerox 914, is introduced— 22 years after it was patented.

1969
Humans land on the moon.

1970
Earth Day is first observed.

1972
The Clean Water Act is enacted to restore and maintain the chemical, physical, and biological integrity of the nation's waters.

1974
Over $5 billion is spent annually on direct mail advertising.

1983
Sales of personal computers in the U.S. have soared from zero in 1975 to $1.4 million; the first Apple Macintosh will be sold in 1984.

1985
Burke Marketing Research concludes that 29% of all office copies made are unnecessary, costing U.S. companies $2.6 billion a year.

Fax machines become commonplace as a means of telecommunications.

1987
The garbage barge *Mobro* sails 5,000 miles in search of a dumping ground; its cargo consists primarily of paper.

1990
Earth Day is celebrated again.

1991
Less than 5% of our nation's original indigenous forests remain.

The Challenge

20

Pulp & Paper North American Factbook: 1990, p. 57.

21

Additional information for this section was compiled from telephone interviews with Jym St. Pierre of The Wilderness Society, and Mitch Lansky, February and July 1991. Mr. Lansky is the author of *Beyond the Beauty Strip: Penetrating the Myths of the Industrial Forest,* a book due out in early 1992 about the environmental problems associated with forest management practices in this country, using the northern Maine woods as a case study.

22

Henry J. Perry, "The Economics of Waste Paper Use," *Pulp & Paper,* Vol. 45, No. 4 (April 1971), p. 83.

The country's annual pulpwood consumption reached 98 million cords in 1989, reflecting a 7% increase over only a three-year period from the 92 million cords used in 1986. Projected demand for the year 2000 is 112 million cords.[20] Yet much attention has been paid to the fact that our growing stock of forest lands has also increased in recent years, and there is likely sufficient timber to supply our expanding paper consumption. But this increase in growing stock is due largely to the planting of tremendously large pulpwood stands in tree farms of the southern United States, based on biologically engineered trees designed for very rapid growth. So while pulpwood supplies exist, the overall management of our forest resources leaves much to be desired. The ongoing destruction of our few remaining magnificent native forests is inexcusable. The environmental problems associated with clear-cutting, extensive aerial spraying of chemicals, and the loss of biological diversity all need to be addressed. If paper recycling rates increase significantly enough, overall pulpwood consumption will start to decrease. This will create an opportunity to improve forest management practices and to preserve what little native forest may be left.[21]

Wastepaper Utilization in the Paper Industry

Recycling has, at times in our national history, played a much more important role than it does today. During both world wars and the Depression, American citizens and industry proved very adept at conserving and recycling resources. During World War II unprecedented quantities of used paper were collected and used in manufacturing; in 1944 the industry-wide utilization rate for wastepaper was 37% of total paper production.[22]

Wastepaper utilization rates are calculated by dividing the weight of wastepaper consumed in paper manufacturing by the total weight of paper products produced. This gives a rough approximation of how much of the total content is supplied by recycled fiber. Because of fiber loss during processing of wastepaper, the final weight of the recycled pulp obtained will be somewhat less than the weight of wastepaper originally consumed. Thus the percentage of recycled fiber supplying the final fiber content of the end products will be slightly lower than the actual wastepaper utilization rate.

23
American Paper Institute, *Capacity Survey: 1990*, pp. 24, 26, and American Paper Institute, *1990 Statistics*, p. 29. Updated estimates for 1990 data provided by API staff.

In the decades following World War II the importance of wastepaper as a fiber source declined, and the overall wastepaper utilization rate in the U.S. paper industry hovered around 23% through much of the 1970s. In the latter half of the 1980s, as interest in recycling increased, the rate began to grow slowly but steadily. Currently wastepaper still provides only about one-quarter of the fiber required for the domestic manufacture of all paper products. In 1989 the U.S. paper industry consumed 20.4 million tons of wastepaper for a total annual production of 78.3 million tons of paper products, yielding a 26% wastepaper utilization rate overall. Virgin wood supplied 60 million tons of pulp, or the majority of all fiber used. By 1990 the wastepaper utilization rate was estimated to have grown to 28%.**23**

The importance of wastepaper as a fiber source for the U.S. paper industry was greatest in during World War II when the wastepaper utilization rate reached an all-time high of 37%. Pulp supplies were short, mills needed fiber, and citizen collection programs were considered patriotic duty. This scrap dealer was photographed in New York City in December, 1941. (Photo by Arthur Rothstein, courtesy of the Library of Congress.)

24
Wastepaper utilization data taken from Franklin Associates, Ltd., *Paper Recycling: The View to 1995* (New York: American Paper Institute, February 1990), Table 1-5, with updated estimates for 1990 data provided by API staff.

25
The term deinking, though pronounced dē-inking, is not generally hyphenated. Deinking is the process of removing inks and other printed or applied contaminants from wastepaper in order to reclaim only the cellulose fiber.

Recycled Printing and Writing Papers: A Marketing Fix?

Within the U.S. paper industry the importance of wastepaper as a fiber source for different paper products varies greatly, and the 1990 industry-wide 28% wastepaper utilization rate does not reflect the importance of recycled materials within all manufacturing segments. In 1990 the tissue industry relied on recycled paper most heavily, using it to supply 52% of all its fiber needs. During this same year, paperboard mills achieved a 37% wastepaper utilization rate, while newsprint mills had a 27% rate. *In contrast, despite the fact that printing and writing papers comprise, by weight, almost 30% of all paper and paperboard products produced in the United States, the wastepaper utilization rate in this segment of the paper industry was only 6%.*[24] Furthermore, the majority of the very small amount of wastepaper used in the manufacture of printing and writing papers came from unprinted paper trimming and converting scrap, while wastes that require deinking supplied only a minor portion of the already limited recycled content of these papers.[25]

The wastepaper utilization rate for the manufacture of printing and writing papers has not changed significantly since 1988, and the American Paper Institute projects that it will only reach 7% by 1995. Yet between 1988 and 1991, while the utilization rate in the printing and writing segment of the

Figure 1.6
Wastepaper Utilization in U.S. Paper Production: 1990

By weight, paperboard is the largest category of paper products produced. Printing and writing papers are the second largest. Each segment of the industry relies on wastepaper to a varying degree, and the different utilization rates are shown for each paper category. Over fifty companies are selling more than 400 different grades of printing and writing papers that they label as "recycled." Yet the wastepaper utilization rate for this segment of the paper industry is only 6%. (Data from Franklin Associates, Ltd., and American Paper Institute.)

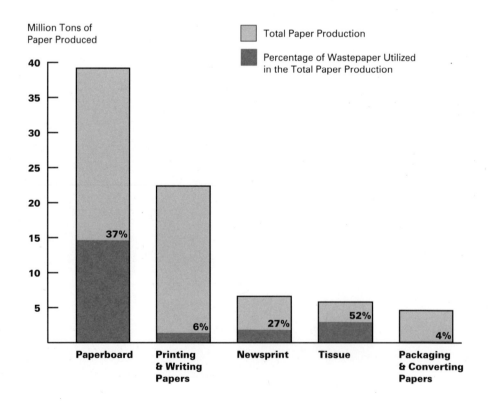

Recycled Papers: The Essential Guide

paper industry remained at 6%, there have been tremendous numbers of new recycled paper grades introduced in the marketplace. There are now over fifty companies marketing recycled papers, and most of these companies sell several different grades. A prominent paper industry consulting firm stated in April 1991 that "one year ago, there were fewer than 80 recycled printing and writing grades on the market. Today there are more than 400!"[26]

It is troubling to purchase recycled paper in the face of such contradictions. Clearly, these facts speak to a need for a meaningful national standard to define a minimum recycled content for papers labeled "recycled." Unfortunately, it also appears that a government body or a responsible, independent private-sector organization is needed to verify the many claims that are being made. Marketing fixes will not solve the real problem— a pressing need for a genuine increase in the recycling and use of wastepaper in all paper products, including printing and writing papers.

Deinking Capacity Expansions in the United States
In 1990 the American Paper Institute, a private trade industry organization whose members include the majority of the U.S. paper producers, announced an industry commitment to expanding wastepaper recycling.[27] Expressed goals are to increase the recovery of available wastepaper in the United States from an amount API estimated at about 30% in 1988 to a total of 40% of all paper wastes by 1995.[28] However, not all of the wastepaper we collect is used domestically; almost one-quarter of it is exported. Further increases in exports will be required to absorb these projected increases in recovery rates, perhaps reaching almost one-third of all wastepaper collected in 1995. In addition, the good news is that many mills in the United States are spending millions of dollars to expand or add capacity to use wastepaper for domestic production of containerboard, newsprint, and tissue products. API projects that by 1995, with these expansions, the overall wastepaper utilization rate in U.S. paper production will reach 30%.

The industry is to be commended for the significant investments it is making in this area. To date, however, virtually all the deinking capacity expansions are limited to products other than printing and writing grades.[29] *Despite the expanded use of wastepaper in newsprint, containerboard, and tissue papers, and given the projected continued growth in paper consumption, even if API's announced 40% recovery goal is met by 1995,*

26
Promotional letter from Jaakko Pöyry Consulting, Inc., selling subscriptions to their new *Recycled Grade-Finder Service* (April 24, 1991).

27
American Paper Institute, press kit including several documents: *News Release; Key Questions and Answers on Paper Recycling; Paper Recycling: The View to 1995* (February 1990).

28
API's 30% and 40% recovery figures include preconsumer wastepaper recovered from paper manufacturing operations, so these figures cannot be compared directly to EPA estimates for recovery of paper from MSW. In 1988 the EPA estimated that only 26% of paper wastes generated were recovered for recycling. (Franklin Associates, Ltd., *Characterization of Municipal Solid Waste in the United States: 1990 Update,* p. 78.)

29
Mary Corbett, "New Deinking Projects Proliferate as Industry Meets Recycling Demand," *Pulp & Paper,* Vol. 65, No. 5 (May 1991), pp. 52–55.

API projects the wastepaper utilization rate for newsprint to grow from the current 27% to 37% by 1995; for paperboard it is expected to grow from 37% to 43%. (Franklin Associates, Ltd., *Paper Recycling: The View to 1995,* Tables 1-5, 1-6, and 1-7; updated 1990 data provided by API staff.)

the total amount of paper requiring disposal in the country will remain essentially the same.[30] Thus, until the printing and writing mills make a greater commitment to expanding their use of recycled fiber through deinking, we will be unlikely to reach paper recycling rates substantial enough to start reducing the total load of paper in the municipal solid waste stream. Nor will we be able to reduce the amount of virgin wood pulp consumed in manufacturing.

There are several deterrents to such an expansion into wastepaper utilization by the printing and writing mills. Compared to recovered paper, virgin wood has provided a relatively easy to use and predictable source of pulp. Most manufacturers are not eager to tackle the uncertainties and challenges associated with increased dependence on this new fiber source. Questions have also been raised by producers about whether customers are really committed to buying recycled printing and writing papers, and whether this demand will justify the costly expansions required to install deinking capacity at these mills. The financial investment needed to build a full-scale deinking plant depends on many factors, including capacity, but it can easily reach $30 million or more for a typical mid-sized system.[31] In addition, despite the growing use of wastepaper as a fiber source, the paper industry is still in the midst of an expansion of its virgin wood pulping capacity that began in the latter half of the 1980s. From 1989 to 1992 virgin wood pulping capacity is expected to grow 5.6 million tons to a total of 70 million tons, an increase of 9% over only three years. Most of this increase is in the production of bleached chemical pulp destined for printing and writing papers.[32] Having already committed to these investments, which give the industry more than enough capacity to meet current demand for pulp production, most mill executives are not anxious to undertake new ones. In light of the country's need to increase wastepaper utilization, it is unfortunate that this expansion in pulping facilities was not directed toward increased deinking capacity instead.

The Power of the Paper Specifier
The deterrents to the increased use of wastepaper in printing and writing papers do not make the necessity of this change less urgent. Establishing even greater increases in paper recycling rates is an essential part of the solution to our environmental problems. If the paper mills are not eager to lead the way, the question is whether economic pressure— the creation of a strong demand for recycled products— will convince more printing and

[30] William E. Franklin, "Incineration: Its Role in Recycling," presentation at *Pulp & Paper Wastepaper I Conference,* Chicago, IL (May 1990).

[31] Based on reported expenditures for several systems announced or under consideration, and interviews with mill executives.

[32] *Pulp & Paper North American Factbook: 1990,* pp. 36–43.

33

A detailed market study, *Projected U.S. Demand for Recycled Printing and Writing Paper,* has recently been completed by Andover International Associates. This study estimated the combined federal, state, and local government purchases across the United States in 1990 to be 1.7 million tons, an amount just under 7% of the total annual U.S. consumption of 25 million tons. The majority of these government purchases tend to be concentrated in the uncoated offset and office paper segment of the printing and writing paper industry. Given the concentration of purchases, and the fact that 1.7 million tons of paper is quite a significant amount, government purchasing does play an influential role in promoting markets for recycled products. (Fred D. Iannazzi and Richard Strauss, Andover International Associates, interviews, August 1991.)

writing mills to take the plunge. In fact, the growth of recycling will be much more successfully driven by market demand than by the collection of waste materials for which there is, at present, limited industrial use.

An awareness of the importance of strengthening the markets for recycled products has led the U.S. federal government, the majority of state governments, and some other local authorities to establish purchasing requirements or preference programs for recycled paper. Yet, while government purchases are significant enough to influence the marketplace, they are only a small portion of total paper consumption. *The private sector consumes over 93% of all printing and writing papers sold in the United States.*[33] Thus the private sector marketplace, acting with enough unanimity, has tremendous additional potential to influence the future directions that mills will take. The day-to-day decisions that designers, clients, printers, publishers, corporations, wholesale and retail establishments, and other consumers make, will have a powerful net effect on this issue and can reinforce the mandate that our governments have initiated.

Those of us who wish to make our voices heard can do so emphatically through creative use of our pocketbooks, by specifying recycled paper loudly and clearly enough for mills to get the message that we are committed to using this product and that further investments will be worth their while. Collectively, all of us who buy paper, whether for government or private use, are doing our part— working on enough different jobs to add up to an annual national consumption of 25 million tons. Every single individual decision influences the equation, pro or con, determining the extent to which we can close the loop and make recycling work.

In the absence of a useful national standard, it is more difficult, but not impossible, to make purchases that support the economic viability of recycled products. So many grades are being touted as "recycled" that buyers now need to look well beyond that sometimes meaningless label in order to understand the details and sources from which the recycled content is supplied. In addition to providing an overview of the issues surrounding recycled papers used for printing, this book is intended to help equip paper specifiers with the knowledge they need to make informed and discriminating choices.

2 A Brief History of Papermaking

From the first days of papermaking, and throughout its history, recycled materials have been used extensively to provide a major source of fiber for the paper industry. In Europe and America, cotton and linen rags were for over 700 years virtually the only fibrous material used. The development of commercial methods for extracting cellulose fibers from wood, in order to make pulp, has been relatively recent. This process was begun by the middle of the nineteenth century, and wood pulp has been the dominant source of fiber in Western papermaking for only the last hundred years. The repulping of used paper to make new paper took place well before the invention of wood pulp and has grown in importance since.

The Invention of Paper

Paper was invented in China at about the time of the birth of Christ. A court official, Ts'ai Lun, so successfully advanced papermaking techniques that his work was recognized by the Emperor in 105 AD. Since his achievements were recorded in ancient Chinese literature, he has been the individual most often credited with this invention. Recent discoveries, however, date rudimentary paper fragments found in China to the first and second centuries BC, confirming the logical supposition that papermaking methods were being developed even prior to Ts'ai Lun's time.[34]

The ready availability of means with which to transmit written communications has been crucial to the development of civilization. Humans have employed many kinds of substrates for writing and drawing, especially since the creation of written language. This quest for suitable materials on which to leave marks has led in various times and places to the use of stones, clay tablets, metals, leaves (particularly of palm trees), wood slabs, barks, skins, and many other materials. Papyrus is known to have been

Rag cutters shredding and sorting rags at the Crane Paper Mill, circa 1920. (Photo courtesy of the Crane Museum.)

[34] *On Paper: The History of an Art,* an exhibition at the New York Public Library, December 8, 1990 – March 2, 1991.

35

Although the word *paper* is derived from the Greek and Latin words for papyrus, this is a bit of a misnomer. Writing papyrus is made from the papyrus plant, which once grew prolifically along the Nile River in Egypt but is now almost extinct there. In contrast to paper, which is formed by the maceration of materials into a liquid pulp of small separate fibers, papyrus is constructed from whole strips of the plant. After being cut, split, and wetted, these strips are carefully laid in two layers, with the grain of the layers perpendicular to each other. These layers are then hammered or pressed together to bond into a single sheet. This section of papyrus contains hieratic writing. (Photo courtesy of The Brooklyn Museum.)

36

Most of what is sold today as vellum and parchment is not made from animal skins, but is paper manufactured from wood or cotton pulp. The name "vellum" is applied both to drafting papers (which are opaque papers subsequently processed to be made transparent) and to some high-grade writing papers. "Parchment," too, is now used to describe a variety of papers treated to create certain effects, reminiscent of true parchments.

37

Dard Hunter, *Papermaking: The History and Technique of an Ancient Craft* (New York: Dover Publications, 1978), pp. 48–52.

38

The earliest known text printed on paper is a Buddhist sutra, which has been dated as being made no later than 750 AD.

used by the Egyptians as early as 2200 BC, and it remained in use well into the end of the first millennium, being replaced by paper only after papermaking was imported to Egypt around 800 AD.[35] Parchment and vellum, made by treating, smoothing, stretching, and drying animal skins— specifically of sheep, calves, goats, or lambs— was probably in use as early as 1500 BC.[36] In China, before the invention of paper, strips of bamboo and scrolls of woven silk cloth were used. Bamboo was relatively cumbersome; silk was very costly. It was the need for a more economical, abundant, and easily used writing medium that drove the invention of papermaking.

Ancient Chinese historical documents record the source of fiber for Ts'ai Lun's paper as discarded cloth and rags, the bark of trees, and well-prepared hemp.[37] What was revolutionary about this new writing surface was its method of preparation— requiring the maceration of these materials into separate fibers, which were suspended in a watery pulp prior to formation. It was a simple yet radical invention that quickly began to influence human civilization.

Papermaking spread rapidly throughout China, and by the third century it had become one of the primary writing materials used for books and scrolls in this part of the world. By the fifth century the Chinese had invented printing, using wood blocks, and it is probable that paper was routinely employed for this purpose.[38] By 610 AD the art of papermaking had reached the islands of Japan via the Korean peninsula, then part of China. The Japanese perfected papermaking techniques and soon became renowned for their excellent paper. Papermaking techniques continued to spread throughout Asia over the first millennium, and a wide variety of plant materials were used for fiber sources. These included the bark of several mulberry plants (also known as kōzo), hemp, China grass (also called ramie), rice straw, bamboo, gampi, and others. In addition to this extensive use of plant materials, there is no reason to believe that the original use of recycled silk fabric scraps and old cloth was discontinued.

Old Paper Provides Fiber in Ancient Japan

The first recorded use of wastepaper for making new paper dates from 1031 AD in Japan, although such a technique was likely to have been employed even earlier by the Chinese. In Japan at this time, old documents and papers were repulped for new paper, which was sold in paper shops.

These woodcut illustrations of papermaking in Japan were published in 1798 in Jihei Kunisaki's book Kamisuki Chōhōki (A Handy Guide to Papermaking). *The basic technique of the time had changed little from the earliest days of papermaking— bark was stripped, soaked, and beaten into a pulp. Moulds were dipped into this pulp to form individual sheets of paper which were pasted onto a board to dry in the sun. (Photos courtesy of the New York Public Library, Spencer Collection.)*

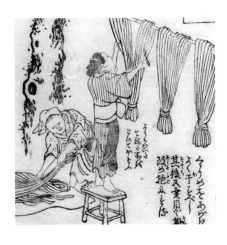

Because bleaching processes had not yet been discovered, the product was grayish in color. This dull colored paper came to be held in very high regard. Even books and manuscripts from the Imperial Library were reported to have been destroyed for use in making this highly esteemed paper. Special papermaking guilds were established for this specific manufacturing art, and demand for this paper continued well into the sixteenth century or even later.[39]

[39] Hunter, *Papermaking,* pp. 54–55.

The Spread of Papermaking to Europe and America

After 700 AD the art of papermaking spread westward from China to Samarkand, then to Baghdad, Damascus, Egypt, and Morocco. The technique did not reach Europe until the twelfth century, and the first European paper was probably manufactured in Spain about 1150 AD. Prior to this time, European scribes and scholars had depended on parchment and vellum for their handwritten manuscripts. Because the process of creating such documents was slow, the demand for new writing materials was small, and the European papermaking industry did not grow especially fast in its first 300 years. However, by the time Gutenberg printed his first Bible in the mid-fifteenth century, there were paper mills in Spain, Italy, France, Germany, Holland, and England. The earliest papers in Europe may not have been looked on favorably compared to parchment or vellum, in part because they were more fragile, but also because the Christian world tended to distrust anything that was imported from Arabic and Jewish civilizations.[40]

Gutenberg's production of movable metal type and the advent of printing in Europe profoundly affected the development of communications, literacy, and, of course, papermaking. As printing spread rapidly over the continent, demand for paper soared, and many new mills sprang up.

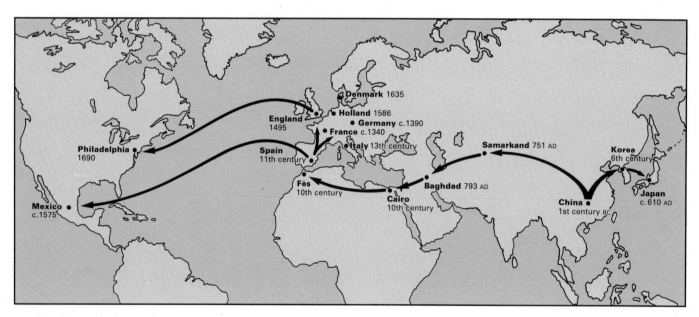

Figure 2.1
The Spread of Papermaking

[40] Ibid., pp. 60–61.

In European papermaking, old rags became the source of fiber, and moulds were made with wire screens instead of the bamboo or reed screens used in China and Japan. The vatman formed each sheet; the coucher laid them down for pressing to remove much of the water. This engraving of a French paper mill is from Art de Faire le Papier *by J. J. Le Français de Lalande, published in Paris in 1761. (Photo courtesy of the Library of Congress.)*

41
In the Western world these paper moulds were made of wire. Originally the moulds were of the laid variety. The "laid-lines" are formed by the impression of the pulp against the closely spaced wires running in one direction, and the "chain-lines" are formed by the less frequently spaced perpendicular wires that actually weave in and out of the laid wires, stitching them together. The wove mould is believed to have been reinvented in Europe around 1750 by either John Baskerville or one of the mills that supplied him with paper. Baskerville wanted to print books on completely smooth sheets of paper. The wove mould, consisting of fine brass screening with closely and equally spaced wires running in both directions, formed such a smooth and even sheet. Both wove and laid moulds had been used in early papermaking in the Orient. But the laid mould had become the type most commonly used and the one originally copied as papermaking moved westward to Europe.

42
From the French *coucher,* to lay down, and the Latin *collocare,* to set in place.

Paper manufacturing rapidly grew into an important industry. European papers, unlike those in Asia, were made exclusively from old cloth rags. At first, linen rags were the primary pulp source, since wool was unsuitable for papermaking. By the middle of the eighteenth century, as imported cotton became more widely available and used for clothing, both linen and cotton rags were used. In the Western world these recycled rags remained virtually the only source of papermaking fiber for well over 700 years, until the mid to late nineteenth century.

Paper during most of this period was still made by hand, essentially in the original manner invented by the Chinese. First, pulp was prepared by macerating the raw materials into individual fibers by mechanical means. Originally, mortars and pestles were used for this purpose, but by 1680 a machine especially suited to beating fiber had been developed. Invented in the Netherlands by a person or persons since forgotten, the machine became known as a Hollander beater, or simply a Hollander. After the pulp was beaten and rinsed, more water was added, and this mixture was put into large vats. A skilled vatman dipped a rectangular paper mould into the vat, lifting it out so as to form a consistent and even layer of pulp over the entire surface area, in order to make even sheets of paper of the same thickness.[41] Water dripped through the moulds, leaving a mat of interlocking fibers. The vatman passed the filled mould to the coucher who laid down or *couched*[42] the sheet of paper onto a drying felt, thus removing it from the mould. A stack of sheets of paper, with a felt between each, was thus made. Moisture was removed by squeezing the stack in a press. Then the paper was hung to dry completely. After drying, the paper was sized by dipping it in a warm glutinous liquid.[43] Then it was pressed and dried again. Its surface was further smoothed by hand burnishing with stone or, later, by pressing hammers.

From Europe, papermaking techniques were imported to the Americas by the Spaniards and the British colonists. The Spaniards established the first North American paper mill in Mexico City around 1575. Despite the fact that the first printing press operating in British Colonial America had been established in Cambridge, Massachusetts, in 1638, it took more than 50 years before the first paper mill was constructed in these colonies. It was erected in 1690, in an area now part of Philadelphia, and operated by William Rittenhouse, who had been a papermaker in Holland. As in Europe, old linen rags were the principal source of fiber. Cotton rags were increasingly employed after the expansion of cotton cultivation and manufacturing in the following century.

Both cotton and linen make excellent raw materials for paper, since both plants are high in cellulose content. The seed hairs of the cotton ball, the portion of the plant used for both cotton fabric and paper, are about 90% cellulose. The flax plant has a cell structure that is about 80% cellulose.[44] A major component of almost all plant cell walls, cellulose is a carbohydrate, and it is these fibers that are extracted from vegetable matter for the essential ingredient of papermaking. This is true regardless of whether the paper is made from wood, bark, straw, and other directly harvested plant material, or from secondary materials such as old paper or cotton and linen rags.

Rag Shortages

As paper consumption increased in Europe and America following the development of printing, there were chronic shortages of rags. In 1666, the English Parliament passed a law forbidding the use of cotton and linen for burying the dead, presumably in order to save these materials for the paper mills. In both England and Germany it became an accepted practice of good citizenry to use only wool to clothe the deceased.[45]

---

[43] Sizing makes papers more impervious to liquids. In the case of printing and writing papers, sizing is needed to prevent the paper from absorbing and blotting the inks. Today most printing papers are made with both internal and external sizing.

[44] James P. Casey, ed., *Pulp and Paper Chemistry and Chemical Technology*, Vol. 1 (2 vols., 1st ed., New York: Interscience Publishers, 1952), p. 1.

[45] Hunter, *Papermaking*, p. 311.

Throughout the eighteenth and well into the nineteenth century, the growing demand for paper, coupled with the shortage of rags, led to steady increases in the price of paper. In America, a great deal of advertising and educational effort was expended by the mills to encourage the habit of saving rags. Newspaper editors and mills proclaimed the habit a patriotic duty. One Massachusetts mill even had a watermark, "SAVE RAGS." Housewives in New England are said to have kept elegant rag bags in their parlors, as part of the furniture, so that every scrap would be saved. In 1776 the Massachusetts House of Representatives passed a decree requiring all towns to appoint an individual whose responsibility was to receive rags for the mills. Tin peddlers bartered for rags, because they could easily sell them to paper manufacturers.[46]

Early Experiments with New Fibers and Deinking
Due in large part to this shortage of rags, experiments with other sources of fiber were begun. The first known suggestion of wood as a pulp source was made by a naturalist, René Antoine Ferchault de Réaumur, in a report to the French Royal Academy in 1719. Réaumur had observed many species of wasps and was struck by their ability to form nests by masticating on wood and subsequently exuding a paste that formed into a durable paper. Despite his extensive writings, Réaumur is not known to have followed his observations with any attempts to make paper. However, over

[46] Ibid., pp. 308–310.

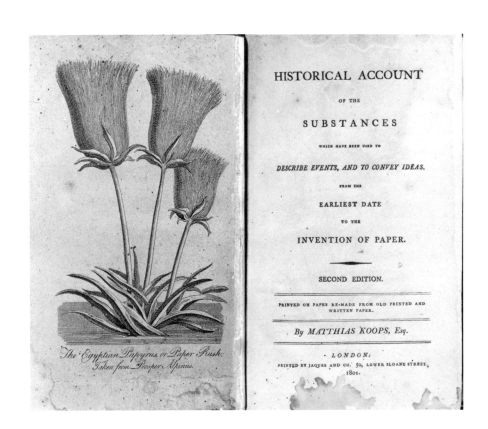

This book, published by Matthias Koops in 1801, was printed on several different papers manufactured at his mill, all of which utilized new sources of fiber. The title page and body of the book are printed on paper made from deinked wastepaper. The frontispiece with the illustration of papyrus is on paper made from straw, and the final few pages of the book use paper made from wood pulp. (Photo courtesy of the New York Public Library, Schlosser Collection on the History of Papermaking.)

the course of the eighteenth century, many others began experiments to extract fibers for papermaking from a wide variety of vegetable matter, including species of wood, swamp moss, potatoes, corn husks, barks, shrubs, and from wasps' nests too. Some experiments with mineral fibers were also made, and a few sheets of paper were even made from asbestos![47]

The first commercial mill in the West to produce paper from materials other than cotton and linen rags was built by Matthias Koops in England in 1801. Koops successfully made paper from a variety of materials, including straw, hay, hemp and flax refuse, wood, bark, and old paper. Koops had obtained several patents from the English patent office, including one granted in 1800— *the first known patent for "extracting printing and writing ink from printed and written paper, and converting the paper from which the ink is extracted into pulp, and making thereof paper fit for writing, printing, and other purposes."*[48] Despite his success with all these new raw materials, Koops's mill was short-lived. He went bankrupt within three years, and old rags continued to be the source of fiber on which papermakers depended.

The Invention of the Paper Machine

This need for new fiber sources was about to be greatly exacerbated. In France, at a paper mill under the direction of François Didot, the first machine for making paper was patented by Nicholas-Louis Robert in 1799, after several years of experimentation. By 1803 an adaptation of Robert's machine had been completed in England, with extensive financing from the Fourdrinier brothers, London stationers who were in need of more paper. Due to problems with their patents, the two brothers never recouped any money from their large investments. Nevertheless, the development of the Fourdrinier paper machine was an amazing technical advance in paper manufacturing. To this day most papermaking machines operate in essentially the same manner and bear the Fourdrinier name. Now paper could be made in continuous rolls. The pulp was first distributed onto a large, continuously moving woven wire screen or mould, with water draining through it. By the time the pulp reached the end of the wire, enough water had been removed to form a sheet. This sheet was transferred via a felt web through pressing cylinders and then wound round a final take-up cylinder roll.

---

[47] Ibid., pp. 313–332.

[48] Ibid., p. 333, citing original English Patent Office documents.

The Evolution of the Paper Machine

Top: *A model of Robert's paper machine.*

Middle: *An engraving of an early Fourdrinier machine, circa 1850.*

Bottom: *A modern Fourdrinier machine that makes publication grade paper in rolls 380" wide at a speed of 3500 feet per minute.*

(Top photo by Laurie Minor courtesy of the Smithsonian Institution; middle photo courtesy of the Smithsonian Institution; bottom photo courtesy of Beloit Corporation.)

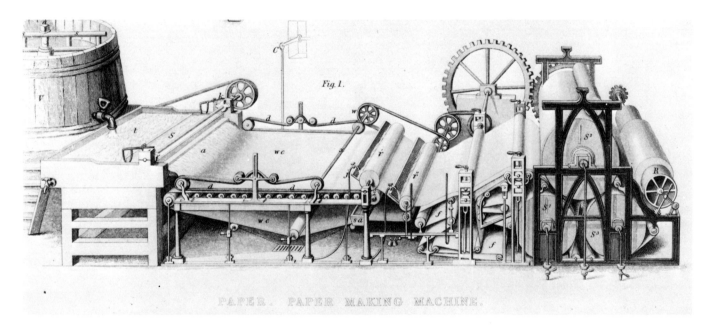

49

Straw has been used as a fiber source since the birth of papermaking in China, where China grass was used. Rice straw has been used in Southeast Asia. In the West, preferred straws for papermaking come from rye and wheat. Although straw is practically no longer used in America, it has regained importance in many European countries and is increasingly being used in many developing countries in Asia, the Mideast, and Central America. As pulpwood supplies become more precious and expensive, the use of straw will probably continue to grow.

50

True manila paper was originally made from hemp, though the "manila" folders sold today in U.S. office supply stores are manufactured with wood pulp. Today hemp plants are used for papermaking in the Philippines and in Central and South America. The use of hemp as a fiber source in this country was spurred by the accidental discovery, in about 1840, of the suitability of old hemp rope by a Massachusetts papermaker desperate for both rag stock and money with which to buy it. Subsequently, larger quantities of hemp and manila were imported to the United States for papermaking. (See Lyman H. Weeks, *A History of Paper Manufacturing in the United States, 1690–1916,* New York: Burt Franklin, 1916, p. 220.)

51

Bagasse is the fibrous residue of sugar cane, left after the harvested plants are crushed to extract the sugar.

The construction of these machines quickly spread to other countries. Paper machines based on this design were soon built in Russia. By 1807 Fourdrinier machines were being manufactured in England for use in mills throughout Europe. Meanwhile, the first paper machine operating in America was based on a slightly different cylinder-type design. It was completed in 1809 at Thomas Gilpin's mill in Delaware, a mill noted for its fine papers because of its cleanliness of manufacturing operations and careful sorting of rags. The first Fourdrinier machine in America was put into operation in 1827 at a mill in Saugerties, New York. It was manufactured in England and shipped here over the objections of English papermakers, who wished to protect their export market for paper. Within two years, other American mills had begun to build and erect their own Fourdrinier machines. Drying cylinders were soon made an integral part of these machines, and paper manufacturing began to accelerate.

Papers from Straw

The search for new fibers now continued in earnest. Prizes were offered for paper made from new materials. In 1828 the first paper in America commercially manufactured from straw was made in Pennsylvania. Beginning in 1830 and continuing for over fifty years, the Columbia County area of New York State had many straw mills in operation and was noted for its straw papers. A few papers were made exclusively from straw, but often some rag fiber was added. Products included many types of papers: newsprint, printing and writing papers, and paperboard. Demand for straw paper grew, and straw mills sprang up in other parts of the United States, eventually including some in the western states where straw was more abundant. By 1871 America was producing over 100 tons of straw paper per day. Straw continued to provide a valuable source of raw fiber material for paper in America and Europe until at least the early 1920s, by which time wood pulp production had greatly expanded.[49]

Continued Growth in Paper Consumption Makes Fiber Shortages Worse

The growth in paper manufacturing and consumption through the nineteenth century was dramatic. In 1800 there were approximately 100 small paper mills in the United States. By 1810 there were approximately 200 mills, all still making paper by hand. By 1860 there were 555 mills in twenty-four states, many making papers by machine. Yet, despite the development of techniques for using straw, the use of hemp for making manila paper,[50] and experiments with corn husks and stalks, bagasse,[51] and many other plant materials, the industry was still primarily dependent on cloth rags for its raw materials.

So the cry for rags continued. Typical of the many advertisements of the day was the heading for this ad: *AMERICANS! Encourage Your Own Manufactories And They Will Improve. Ladies, Save Your RAGS!* [52]

By the mid-1850s rags were in such short supply that several mills are said to have imported whole Egyptian mummies or their cloth wrappings for use of the cloth in papermaking. A paper mill worker in New York State described how the linen cloth, having been tightly wrapped around the bodies, retained its mummy shape and would spring back into this form if one tried to straighten it. Another story, eventually relayed by the mill owner's son, stated that a cholera epidemic was started among the rag pickers and cutters at a mill in Maine, where they cut the cloth from these imported ancient bodies.[53]

Following the Civil War, increases in paper consumption continued as the American continent underwent rapid industrialization. Settlers took possession of the vast western territories, displacing the native inhabitants and establishing a completely different mechanized culture in the process. Many newspaper publishing operations were started in the expanding cities and settlements, and many more books were published. At the same time, another revolution in paper manufacturing was starting— the use of wood as a fiber source.

The Development of Wood Pulp

Wood is not a perfect source of fiber by any means. Species vary, but the cellulose content of wood is roughly only 50% by weight, much less than that of linen and cotton. In addition, wood contains lignin, a binding material within the cell structure, which may be as much as 25 or 30% of the total weight. If the lignin is not removed from the wood pulp, the resulting paper is weak and deteriorates fairly rapidly; light causes it to yellow. Nevertheless, there are two principal reasons why wood has become the preferred raw material used for Western papermaking in the last century. First, the available supply, especially in areas such as the United States, Canada, and Scandanavia, has, until recently, been abundant and relatively inexpensive. Second, the use of large quantities of virgin material has given manufacturers a more consistent pulp source— one that is easier to work with because it is free from the labor-intensive and technical problems associated with the sorting of recovered materials and the removal of contaminants that is required with old rags or used paper.

52
This ad was published in *The Pittsfield Sun* in 1801 by the three founders of the first paper mill in Dalton, Massachusetts. Eventually owned solely by the Crane family, the Crane Paper Company is the only company left in the United States that manufactures paper primarily from cotton fibers. It is also known for its production of currency papers for the U.S. and other governments. (Photo courtesy of the Crane Museum.)

53
Hunter, *Papermaking*, pp. 382–384.

[54] Ibid., pp. 376–381, 550–568.

[55] Weeks, *A History of Paper Manufacturing in the United States,* pp. 286–297.

[56] Hunter, *Papermaking,* pp. 389–394, 561–581; Edwin Sutermeister, *The Story of Papermaking* (Boston: S.D. Warren Co., 1954), pp. 59–100.

The Groundwood Process

The first commercially viable method for making wood pulp employed the groundwood pulping process. This was developed in Germany in the 1840s and, independently, in Nova Scotia at about the same time. Groundwood pulp, as the name implies, is simply wood that has been ground up in such a way as to retain the fiber. American newspapers started printing on groundwood papers in the 1860s, and by 1880 many were using them exclusively.[54] The discovery of methods for using wood as a fiber supply had a dramatic effect on paper markets; after years of escalating prices, the cost of paper finally started coming down. The cost of newsprint in the United States in the 1860s was about 25¢ per pound, but by 1897 it had fallen as low as 2¢.[55]

Chemical Wood Pulps

While the groundwood process yielded large quantities of cheap paper, the quality was greatly inferior to that of rag papers. So the search continued for chemical processes of pulping wood in order to remove the lignin, extract a purer cellulose fiber content, and thus produce papers that would be more durable and of better quality for printing. Three principal methods for chemically pulping wood were eventually developed; all involve cooking wood chips with a variety of chemicals under specific conditions of temperature and pressure. The first was the soda process, developed in England in 1851. It was used by a U.S. paper mill as early as 1855, and its importance in wood pulp manufacturing grew during the latter half of the nineteenth century. This pulping method, however, has since been superseded by other methods, and is currently little used.

The second chemical pulping process developed, the sulfite process, was an acidic method perfected in Sweden in the 1870s. The first commercial sulfite pulp was made in the United States in 1882, and sulfite pulps are still commonly made. The third chemical pulping method, the sulfate process, uses alkaline compounds in the cooking process. Also known as the kraft process, it was developed in Germany in 1883. However, kraft pulp was not made in North America until after the turn of the century. It was manufactured in Canada in 1907, and two years later sulfate pulp was first made in the United States.[56]

These three wood pulping methods, along with some newer hybrid techniques, are still used to manufacture most of the paper we consume today. Groundwood pulp is used for newsprint as well as a few other papers.

Of the chemical processes, it is the sulfate or kraft process that is most widely used; about 70% of all wood pulp in North America is produced by this method. The process was named from the German word *kraft*, meaning strong.[57] It yields a pulp with very high strength properties, especially compared to groundwood pulps. This method also has the advantage that it can utilize a wide range of wood species. As is true of chemical wood pulps generally, however, the final fiber yield is very low, only about 40–55% of the wood by weight.[58]

Starting in the latter half of the nineteenth century, with the advent of methods for pulping wood, paper could be produced much more quickly and cheaply. Finally, an abundant source of fiber was readily available. Paper machines continually got bigger and faster. Originally running at speeds of 100 feet per minute, they had reached 650 feet per minute by the early part of this century.[59] By 1899, only 90 years after the paper machine was first introduced in this country, there were 1,232 paper machines operating in over 750 different mills.[60] After 1900 U.S. paper manufacturing and consumption began to grow at an unprecedented rate.

Recycled and Deinked Wastepaper Rediscovered

In the early twentieth century the United States continued to import rags and some wastepaper, in order to supply a portion of its pulp needs. A drop in these imports at the start of World War I contributed to significant shortages, and manufacturers cried out for fiber. By 1916 the public was once again being exhorted to save old rags. Only this time, the pleas included a new request— for old paper as well. The U.S. Department of Commerce, under Secretary William C. Redfield, advertised extensively on behalf of paper mills to encourage the saving of rags and wastepaper. Used paper became a valuable and salable commodity. For the first time in America, thousands of tons of old books, newspapers, and business papers were recycled by the mills.[61]

The deinking of wastepaper for pulp, at least in significant quantities, probably commenced at about this time. Dr. Thomas Jasperson, after years of research, successfully worked with several mills in the midwest to make paper from the deinking of wastepaper, and he obtained a U.S. patent for his technique.[62] Two of the printing and writing mills currently operating deinking systems report beginning those operations around 1915 or close to the turn of the century.

---

[57] Technically, kraft paper is any paper made from sulfate pulp. The "kraft" paper in the stationer's store is simply unbleached paper, presumably made by the sulfate process.

[58] James P. Casey, ed., *Pulp and Paper Chemistry and Chemical Technology*, Vol. 1 (4 vols., 3rd ed., New York: John Wiley & Sons, 1980), pp. 162–165.

[59] Today some paper machines run at speeds over 3,500 feet per minute. The width of paper made on these machines has continued to increase to as much as 400 inches.

[60] Weeks, *A History of Paper Manufacturing in the United States,* pp. 290, 296.

[61] Ibid., pp. 286, 319. See also the series of reports on Secretary Redfield's efforts that appeared in many issues of *The New York Times,* including "Magazines Watch the Paper Supply" (March 14, 1916), "Pleased at Rag Saving" (April 3, 1916), and "Shortage of Paper Calls for Economy" (June 17, 1916).

[62] "New Papers From Old," *The New York Times* (November 13, 1916).

In the early part of this century American paper mills imported rags and wastepaper from Europe to help supply fiber to the paper industry. When France and other European countries placed embargoes on their exports of these materials in 1916, U.S. manufacturers, already faced with a growing demand for paper, became very short of fiber. The paper supply became quite limited and prices rose. On behalf of the mills, the Department of Commerce advertised extensively to encourage the saving of wastepaper and rags. Well over one million of these fliers were distributed and posted in public places. The department also published pamphlets with specific instructions about what types of paper to save and how to sort it. (Photo courtesy of the National Archives.)

> *Please post in a conspicuous place.*
>
> ### DEPARTMENT OF COMMERCE
> ### : : : WASHINGTON, D. C. : : :
>
> # SHORTAGE OF PAPER MATERIAL
> ## *Save Your Waste Paper and Rags*
>
> The attention of the Department of Commerce is called, by the president of a large paper manufacturing company, to the fact that there is a serious shortage of raw material for the manufacture of paper, including rags and old papers. He urges that the Department should make it known that the collecting and saving of rags and old papers would greatly better existing conditions for American manufacturers.
>
> Something like 15,000 tons of different kinds of paper and paper board are manufactured every day in the United States and a large proportion of this, after it has served its purpose, could be used over again in some class of paper. A large part of it, however, is either burned or otherwise wasted. This, of course, has to be replaced by new materials. In the early history of the paper industry publicity was given to the importance of saving rags. It is of scarcely less importance now. The Department of Commerce is glad to bring this matter to the attention of the public in the hope that practical results may flow from it. A little attention to the saving of rags and old papers will mean genuine relief to our paper industry and a diminishing drain upon our sources of supply for new materials.
>
> A list of dealers in paper stocks can be obtained from the local Chamber of Commerce or Board of Trade.
>
> WILLIAM C. REDFIELD, Secretary.
>
> WASHINGTON : GOVERNMENT PRINTING OFFICE : 1918

63
Henry J. Perry, "The Economics of Waste Paper Use," *Pulp & Paper*, Vol. 45, No. 4 (April 1971), p. 83. This article cites the wastepaper utilization rate for "book paper" over intervals of several years. It was 19% in 1947, increasing to 25% in 1954, and staying at that level until 1958.

The use of wastepaper, especially in printing and writing grades, progressed slowly. The rapid development of wood pulping capabilities seemed to obviate much need to save or reuse existing resources, and the use of virgin pulping methods has dominated this century. Wastepaper, however, increased a great deal in importance as a fiber source during World War II, and the utilization of wastepaper in printing and writing grades peaked at over 20% in the 1950s— a rate far higher than the 6% of today.[63] As resources became relatively cheap and abundant in the decades following the war, the importance of wastepaper again waned. Little has been recorded about the changes in deinking capacity during this century, although even until the mid-1970s there were still a fair

64
Jobe B. Morrison, Miami Paper Company, and Donald Meyers, Butler Paper Company, telephone interviews, March 1991.

number of printing and writing mills using deinking systems. Most of these mills were small operations that had started using wastepaper for economic reasons (a cheaper source of fiber) or because of limited virgin pulp supplies during World War II. Many were only marginally viable operations that had difficulty competing with the increasingly large corporate giants who operated their own timberlands, pulping operations, and related timber products divisions. In the 1970s, when the United States adopted stricter environmental laws, these small mills did not have the extensive capital needed to install the necessary pollution control equipment. And unlike the larger corporations, which could better absorb these costs, they closed.[64] Thus, despite the current interest in recycling, there is much less deinking going on today in the printing and writing segment of the paper industry than there was twenty or forty years ago.

3 How Recycled Papers Are Made

Recycled printing and writing papers are made by repulping a variety of high-grade paper wastes to reclaim only the cellulose fiber and then using this old fiber in the manufacture of new paper. There are many different recycled papers on the market, and there are no standard formulas for the types of wastes used or the total percentage of recycled content included. Currently, the major ingredient contributing to the "recycled" content of many papers is unprinted trimming and converting scrap from the paper mills themselves, as well as from independent envelope converters or other manufacturers. In the United States the percentage recycled content most often refers to the percentage of the total fiber content that is recycled. Typically this may be 50%, and only a very few papers are made with a 100% recycled fiber content. If printed paper wastes are used, they are customarily, though not always, deinked prior to their inclusion in the new paper. There are only a few papers made in which a high percentage of deinked pulp supplies the recycled content, but they are available. Contrary to common assumption, almost none of the office paper collected in the United States, if it contains photocopied or laser-printed materials, is recycled into new printing and writing papers.

Deinked pulp made from 100% wastepaper being pumped into the holding tank at Cross Pointe's Miami Paper Mill. In appearance, there is nothing that readily distinguishes this pulp from virgin pulp. (Photo © Claudia Thompson.)

Wastepaper: A Diverse Resource

Understanding the wastepaper stream requires a good working knowledge of pulp manufacturing methods, the tremendous variety and types of paper made, the subsequent printing and finishing processes that may be applied, and the appropriateness of the many resulting paper products for collection and recycling into new products. This knowledge is the province of specialists who work as buyers of wastepaper for the paper mills, of dealers and brokers who buy and sell these wastes, and of the growing body of people responsible for setting up and running collection programs in order to separate these wastes into recoverable and marketable commodities.

There are close to one hundred grades of wastepaper bought and sold in the United States. The Paper Stock Institute of the Institute for Scrap Recycling Industries (ISRI), a trade industry group representing waste dealers and brokers, publishes a listing of 51 standard grades and 33 specialty paper stock grades that they recognize for transactions in the United States and Canada.[65] Allowable parameters of recovered paper are defined for each of these grades. These standards serve as a starting point for establishing the characteristics of the wide variety of wastepaper available, but it is not uncommon for buyers, sellers, and dealers to recognize their own specialty grades, which may differ somewhat from the ISRI guidelines.

Five General Categories of Paper Wastes

This myriad of wastepaper grades is grouped by the Department of Commerce and the paper industry into five general categories. Each category consists of grades that share certain general characteristics and differs from other categories in its suitability for recycling into the various types of new paper and paperboard products. The five categories are explained below.

Pulp Substitutes: The highest quality of wastepaper available, pulp substitutes consist of completely unprinted paper scrap— most commonly from bleached white papers. These wastes are generated by the mills themselves, in the form of damaged rolls, obsolete inventories, and trimmings from the converting of huge finished paper rolls into smaller web rolls, cut sheet sizes, envelopes, and so forth. These wastes are also created by independent envelope converters and finishing houses as well as other manufacturers or printers that trim excess paper material prior to use. Pulp substitutes are exactly what their name implies: they are a substitute for pulp and may be thrown directly into a paper mill's beater for repulping and immediate manufacture into new paper.[66] Pulp substitutes are the most expensive grade of wastepaper, occasionally approaching the price of some bleached virgin wood pulps. In late 1989 and early 1990 the grade known as "hard white envelope cuttings," used for many products including printing and writing papers, was selling for well over $400 per ton, approaching the cost of bleached hardwood pulp. Prices have since dropped due to competition from an increasing supply of virgin pulps, but this grade has consistently sold in the $300–400 per ton range.[67]

---

[65] Paper Stock Institute, *Scrap Specifications Circular 1990: Guidelines for Paper Stock: PS-90 Domestic Transactions* (Washington, DC: Institute for Scrap Recycling Industries, 1990).

[66] Mills that manufacture paper do not always produce their own pulp. Some mills buy market pulp from other mills that are either sole producers of pulp or produce it in excess of their own capacity to use it. Market pulp is usually dried into rough sheets, which are stacked and cut into a uniform size and put into bales wrapped with a covering sheet of paper and baling wire for shipping. In paper manufacturing it is the original production of wood pulp, rather than the final production of the paper, that requires the most energy and generates most of the associated pollution.

[67] Selling prices for wastepaper, both domestic and export, are reported extensively in the monthly periodical *Paper Recycler*, published by Miller Freeman Publications. The first date of issue was October 1990; see this issue and subsequent issues for a monthly tracking of prices in the United States.

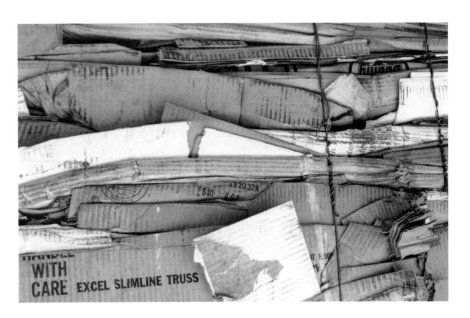

Pulp substitutes (above and top right), consisting of both unprinted paper trimmings and discarded rolls of paper, are being used by a tissue mill to make napkins, towelling, and other tissue products. High-grade deinking wastes (middle right) are also being used by this same tissue mill. Old corrugated containers (bottom right) are collected for recycling into new containerboard. (Photos © Susan Larocque.)

High-Grade Deinking: The grades in this category are the next highest quality of paper wastes available. They are readily used by any mill with deinking capacity, and they generally command a fairly good price per ton— often in the range of $100–200 and sometimes even higher. High-grade deinking wastes consist of bleached paper stock, usually white, that has some printing, though generally without heavy ink coverage. These wastes are generated by offset printing plants as well as other manufacturers, converters, and finishing houses producing envelopes, boxes, folders, and similar products. This wastepaper category also includes some computer print-out papers collected from data processing centers or offices. However, this category does not include most office wastepaper, particularly if it contains laser-printed or photocopied materials.

Old newsprint is the type of paper most frequently collected in residential recycling programs. (Photo © Susan Larocque.)

Mixed: These wastes include the broadest range of grades, a very diverse group of materials that are hard to classify elsewhere. Most of the wastes generated by office collection programs fall into this category, as do similar wastes generated by retail stores and some residential collection programs. The category also includes many wastes generated by printers and manufacturers that do not qualify as high-grade deinking. It includes both white and colored papers, coated papers, and magazines. Mixed grades commonly sell for somewhere between zero and $50 per ton, with the usual price toward the lower end of the range.

News: This is one of the easiest categories to recognize and classify. It consists of several grades of newspaper wastes, including old newspapers collected from residences or offices, and both printed and unprinted trimmings and over-issues from newspaper printing plants. In the industry, these grades are commonly referred to as ONP (old newsprint). Markets for ONP have fallen drastically or collapsed in the past few years as collection efforts have surged far ahead of the industry's capacity to utilize this fiber. Several newsprint deinking expansions are underway, which should improve this situation somewhat.

Corrugated: Also an easy category to recognize and classify, corrugated grades include used boxes collected from offices, stores, and residences, and the cuttings generated during the manufacture of boxboard and corrugated containers. Brown kraft grocery bag cuttings and similar materials are also included. The acronym widely applied to this group is OCC (old corrugated containers). OCC generally sells for less than $50 per ton.

Recycled into What Form?

In general, paper wastes are recycled either into equivalent products or else "down the ladder" into new paper that is lower in grade or more easily manufactured. Pulp substitutes and high-grade deinking wastes are the primary types of wastepaper currently being used in the United States for the recycled content of printing and writing papers. These two grades of wastepaper are also widely used to make many other types of paper products, such as tissue, paperboard, containerboard, packaging papers, and even the paper facings of wallboard building materials.

Few mixed paper wastes are used in the manufacture of recycled printing and writing papers. They are consumed primarily in the production of corrugated containers, paperboard, and construction paper, and some mixed wastes are used in tissue production. Both mixed wastes and ONP are also used to make some molded paper products such as egg cartons and cushioning for packaging containers. Old newsprint most commonly is recycled into new newsprint; it is also used to make paperboard and folding cartons. As new and better deinking technology for newsprint is coming on line, old magazines (OMG) are increasingly being incorporated, in small quantities, into recycled newsprint. Corrugated paper wastes are almost always recycled into new containerboard or paperboard.

The use of lower grades of paper to make higher grades of paper, sometimes called "upcycling," occurs rarely. For example, newsprint and old magazines are occasionally employed in the manufacture of printing and writing papers. This is the exception rather than the norm, however, and has been done for only a very few grades.

Tissue products— napkins, paper towels, toilet tissue, and so on— are nonrecoverable wastes. Thus there is essentially no use of any tissue wastes for new paper. Recycled fiber is used quite extensively, however, for the production of new tissue products. Overall about one-half of the fiber used in this segment of the industry comes from wastepaper. A large percentage of the wastepaper content is directed toward tissue products that are sold commercially to large institutional buyers rather than to the consumer tissue products sold in stores and supermarkets.[68] Pulp substitutes and high-grade deinking wastes provide the majority of the wastepaper used by this segment of industry, but ONP and mixed grades of paper wastes are recycled into tissue as well.

68
Tissue manufacturers still perceive a strong consumer market for the brightest, whitest, and softest tissue products, and virgin fiber is used almost exclusively for the production of these "premium" consumer grades. Economy tissue grades, even though they may not be labeled as having recycled content, are more likely to utilize wastepaper for at least some portion of their fiber. (Jeanne Carroll, "Outlook for the U.S. Tissue Market," paper presented at *Pulp & Paper Wastepaper I Conference,* Chicago, IL, May 1991, and Jerry Goodman, "Secondary Fiber to the Tissue Industry," paper presented at *Fifth Annual International Papermaker's Tissue Conference,* Toronto, Ontario, August 1990.)

Wastepaper Collection and Utilization

In the United States we now collect between 26 and 30% of all available paper wastes for recycling. The EPA estimates that we recovered 26% of the paper in the MSW stream in 1988. The American Paper Institute reports an even higher recovery rate for that same year— 30% of paper consumption. The API's estimate includes preconsumer paper trimming and converting scrap recovered from manufacturing operations while the EPA does not consider such wastes to be part of the MSW stream.[69]

As the world's largest exporter of wastepaper, we rely on foreign buyers to purchase a significant portion of what we collect. In 1990 we exported 22% of the 28.9 million tons of paper and paperboard wastes recovered— to Pacific Rim countries such as Korea, Japan, and Taiwan, to Mexico and Canada, and to Spain, Italy, and other European countries.[70]

Of the 22 million tons of wastepaper remaining after export in 1990, paperboard and tissue production consumed the lion's share— an astounding 17.7 million tons. Collectively, exports and the domestic manufacture of paperboard and tissue products accounted for 83% of all the wastepaper recovered for recycling in the United States. Only 5% of all the wastepaper collected was used for the domestic manufacture of printing and writing papers, even though these papers are second only to paperboard in terms of total production and represent almost 30% (by weight) of all the paper products produced in this country.

[69] Franklin Associates, Ltd., *Paper Recycling: The View to 1995* and *Characterization of Municipal Solid Waste in the United States: 1990 Update.* 1988 is the last year for which complete recovery statistics have been published by both EPA and API.

[70] Franklin Associates, Ltd., *Paper Recycling: The View to 1995,* Table 1-5, with updated 1990 estimates provided by API staff. *Pulp & Paper North American Factbook,* 1990, pp. 314–316. *Paper Recycler,* Vols. 1 and 2.

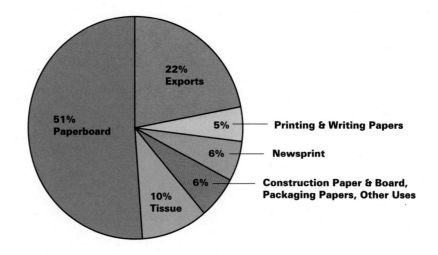

Figure 3.1
Exports and Domestic Uses of Recovered Wastepaper: 1990

Paperboard and tissue products consume four-fifths of the wastepaper remaining in the U.S. after export. Even though printing and writing papers represent almost 30% of all paper products produced, only 5% of all recovered paper wastes are used for their manufacture. (Data from Franklin Associates, Ltd., and American Paper Institute.)

Figure 3.2
Wastepaper Utilization in Printing and Writing Papers

In 1990, only 1.4 million tons of wastepaper were consumed for the total U.S. production of 22.4 million tons of printing and writing papers, yielding a 6% wastepaper utilization rate. Pulp substitutes provided the majority of the recycled fiber used, and high-grade deinking and other wastepaper grades provided only 2% of all fiber used.

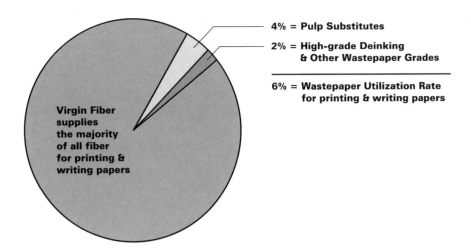

71
Franklin Associates, Ltd., *Paper Recycling: The View to 1995,* Table 1-5, with updated statistics for 1989 and 1990 provided by API staff.

72
Ibid. Over the three years, the percentage of pulp substitutes destined for printing and writing papers ranged from 29 to 31%; the percentage of high-grade deinking wastes used for these papers ranged from 12 to 12.5%.

Wastepaper Utilization in Printing and Writing Papers
The minor role of wastepaper as a fiber source for printing and writing mills is also clearly reflected by the wastepaper utilization rate. In 1990, only 1.4 million tons of wastepaper were used for the entire annual domestic production of 22.4 million tons of printing and writing papers. Moreover, of this relatively small amount of wastepaper used to supply recycled content, it was pulp substitutes rather than deinking wastes that provided the majority of the recovered fiber. Over two-thirds of the wastepaper, or almost 1 million tons, came from pulp substitutes; less than one-third, or only 450 thousand tons, consisted of deinking or other grades. *Thus the overall wastepaper utilization rate for printing and writing papers was only 6%. Furthermore, deinking wastes provided only 2% of all fiber used for the total U.S. production of printing and writing papers.*[71]

Much higher percentages of recovered fiber could be achieved in printing and writing paper production if more of both the pulp substitutes and high-grade deinking wastes currently being recovered were directed toward these products. Yet, despite being the highest grades of wastepaper available and well-suited for inclusion in printing and writing papers, the majority of these wastes are instead sold for domestic paperboard and tissue production or exported. In each of the three years from 1988 through 1990, only about 30% of all pulp substitutes and 12% of the high-grade deinking wastes collected were used for printing and writing papers.[72] In addition, while almost all available pulp substitutes are already recovered, the paper industry estimated that in 1988 only about 37% of the available high-grade deinking wastes were collected to begin with; improved recovery will make more of these wastes available.

Figure 3.3
Recycling of Pulp Substitutes: 1990

Three million tons of pulp substitutes were recovered for the variety of uses shown here. The paper industry estimates that virtually all of the available pulp substitutes are being recovered.

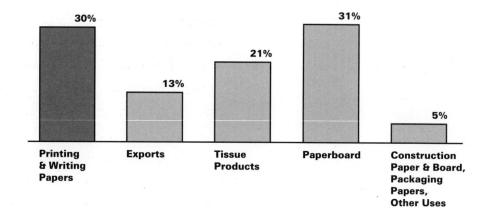

Figure 3.4
Recycling of High-grade Deinking Wastes: 1990

Just under three million tons of high-grade deinking wastes were recycled for the uses shown here. More of these wastes are available; the estimated recovery of high-grade deinking wastes was well under one-half of the available supply.

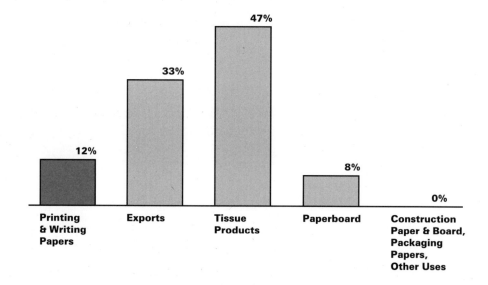

73
Forty-nine U.S. manufacturers are listed in the Jaakko Pöyry, *Recycled Grade Finder*, Vol. 1, No. 2 (August 19, 1991). Several additional American companies not listed here are listed as manufacturers of recycled paper in other sources.

74
The section *Recycled Printing and Writing Papers: A Marketing Fix?*, in chapter 1, discussed wastepaper utilization rates and the marketing of recycled papers.

Composition of Recycled Printing and Writing Papers

Currently more than 50 different paper companies in the United States are producing some grades of printing and writing paper that they classify as recycled.[73] There are, however, no uniform national standards that must be met by paper producers using this classification to market their papers. Given the wastepaper utilization rates, it is clear that pulp substitutes supply at least two-thirds of the recycled fiber content for these papers. In addition, it is reasonable to conclude that internal mill wastes are supplying, on an industry-wide basis, the most substantial portion of the fiber content now being counted as recycled. How else could the wastepaper utilization rates remain almost constant over a several-year period during which the number of papers being sold as recycled has increased at least fivefold?[74]

75

This change has occurred since the introduction in 1988 of EPA procurement guidelines for the recycled content of printing and writing papers. These guidelines suggest a minimum 50% recycled fiber content, and EPA definitions allow many dry paper wastes generated by the mill to be counted toward recycled content. Most paper mills, on questioning, will readily acknowledge using internal mill wastes to supply at least some of the wastepaper content for their recycled grades. At least one mill salesperson visiting my office to promote a "new" recycled paper has literally said that their company now saves up enough of their mill wastes for use in that particular grade, in order to call it recycled, simply in the hopes of selling more paper. On several occasions I have heard paper distributors jokingly refer to some printing and writing papers as "EPA Recycled." See chapter 4 for a discussion of the EPA guidelines.

76

An integrated paper mill produces all of its own pulp and then produces paper directly on the same site. It is a huge operation in which virgin logs or wood chips are put into one end of the plant and finished paper is rolled off paper machines at the other end, usually some distance away.

Wastepaper from the paper mills that is used to supply recycled content includes trimming scrap generated as the large finished rolls are taken off the end of the paper machine and converted into smaller web rolls and cut sheets. Damaged rolls and obsolete inventories of old paper are also repulped, and many mills count this fiber toward the recycled content in certain grades. These internal mill manufacturing wastes have always been reused by the paper companies because it has always made economic sense to do so. Traditionally, this scrap was reused across the board, providing a small portion of the fiber in most grades made by the mill. But it has now become common practice for mills to save up their own internal scrap in order to use it in only one or a few grades that they now classify as recycled.[75]

Not all pulp substitutes used in papermaking are generated by the mills themselves, however. Similar scrap is produced by independent envelope converters, printers, and graphic arts finishing houses, which trim unprinted paper sheets down in the process of producing their products. Some producers of recycled paper use a combination of internal mill scrap and paper wastes purchased from outside dealers and brokers. Historically, there has been a good market for pulp substitutes for many years, as they are a valuable source of fiber. Indeed, some of the smaller, non-integrated paper mills have found it necessary to utilize fiber sources such as these, which cost less than virgin pulp, in order to survive and compete with the fully integrated "supermills" run by huge corporations that have increasingly dominated the paper business.[76] For example, the French Paper Company in Niles, Michigan, began using pulp substitutes purchased from outside dealers and brokers as a principal source of fiber in the 1970s, because the economics were favorable.

Many of the mills that were using wastepaper in the manufacture of new paper did not market their papers as recycled at first. There was a concern that potential customers might be leery of papers made with reclaimed fiber. Instead, these mills quietly went about their business, making fine papers that competed with virgin paper, and most purchasers were unaware of the manufacturing specifics. By the mid-1980s, though, as interest in recycling began to grow, the French Paper Company and several other mills that deink wastepaper— the P.H. Glatfelter Company, the Miami Paper Company, and a few others— were becoming more widely known as producers of recycled papers.

Recycled Content Percentages

There is much more to paper than meets the eye. Printing and writing papers are made from a combination of different materials. Cellulose fiber, the essential ingredient necessary to form the sheet, generally comprises the principal and heaviest component of the paper. Internal fillers— principally clay, calcium carbonate, and sometimes titanium dioxide— are added to the fiber for a variety of reasons. Calcium carbonate adds opacity and brightness to the sheet and provides a suitable buffering agent for alkaline papers. Clay and other fillers can also substitute for fiber and cut down production costs. Titanium dioxide is an optical brightener and opacifier, added to make the brightest, whitest sheets. Fluorescent dyes are sometimes used as well, to make papers even brighter. In order to make the paper more impervious to inks, internal sizings and surface coatings or sizings are also used.

This filler content of paper varies from mill to mill and from grade to grade within a mill. It may be as little as 5% of the paper weight in some of the finest papers, or as much as 28% in some uncoated papers.[77] Typically, filler will comprise at least 12% of the paper, and it often represents an even larger portion of the total weight. Filler is sometimes referred to as "ash content." If the paper is burned, the organic cellulose fibers combust, and the ash remaining is the noncombustible portion of the paper— the clays and other inorganic compounds. The total weight of any paper is thus comprised of some small part moisture due to atmospheric humidity (perhaps 5%), a large part fiber (usually 70–90% in uncoated papers), and the remaining part fillers and coatings (roughly 5–25% depending on the paper). If the paper is fully coated on two sides, the clay coatings give it a very high ash content, and the fiber percentage of the sheet would be even smaller— typically under 60–70%.

In speaking about the percentage recycled content, then, one must be careful to specify whether this percentage refers to total fiber content or to total paper weight. These are very different measurements. It is relatively easy to achieve a 50% recycled content if this measurement is made by total fiber content or, as it is also called, total fiber weight.[78] Because fiber content is only one component of paper weight, however, it is more difficult to achieve a recycled content that is 50% of the total paper weight. To do so, more than 50% of the pulp must come from recycled sources.

77

Handmade papers are also often made with the addition of some fillers, sizings, and dyes, though in many handmade papers the use of such additives is kept to a minimum. Unsized papers are known as *waterleaf* papers; these papers, such as blotting, filter, and toweling papers, are very absorbent.

78

The matter of definitions is further confused by the fact that "total weight" can be interpreted to mean either total paper weight or total fiber weight. Thus it is essential to be precise about exactly what is specified. Buyers should avoid use of the term "by weight," since without the modifier (paper or fiber) it does not specify which measurement is, in fact, being used.

[79] The Canadian EcoLogo program is discussed in chapter 4.

For example, if the paper has a filler and moisture content that is 25% of the paper weight, then only 75% of the total paper weight comes from fiber to begin with. To meet a standard that defines recycled content as 50% of the total paper weight, at least two-thirds of the fiber would need to come from recycled sources. If the paper has an even higher filler content— such as in a fully coated glossy paper— the proportion of the paper weight that is fiber would be even less than 75%, and the proportion of the fiber that is recycled would have to be greater than two-thirds.

Mills, vendors, and the public often refer to the recycled percentage of a particular paper without clarifying whether this percentage is based on total fiber content or total paper weight. The general custom, in the United States, has been for producers to use percentages that refer to fiber content rather than paper weight. However, some jurisdictions in the United States, such as California, require recycled percentages measured by total paper weight. The Canadian labeling standards, established through the EcoLogo program, also require a minimum recycled content as a percentage of total paper weight.[79]

Figure 3.5
Recycled Content Percentages

The same paper requires a different formulation to meet a 50% recycled standard, depending on whether the standard is defined by total fiber content or total paper weight. This chart compares the two situations. In definition A, recycled content is defined only in reference to the fiber portion of the paper. In definition B, recycled content is defined with respect to all components of the paper, the total paper weight. For the sake of comparison, the percentages shown within the pie in both situations are calculated based on total paper weight. If the paper is 50% recycled by total fiber content, instead of by total paper weight, substantially less recycled fiber is required to meet the standard.

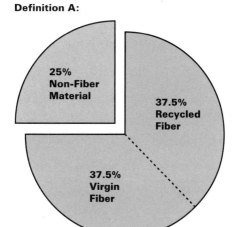

Definition A:

50% Recycled By Total Fiber Content
(By total paper weight, this has only a 37.5% recycled content.)

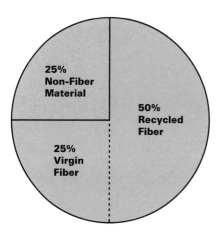

Definition B:

50% Recycled By Total Paper Weight
(By total fiber content, this is two-thirds, or 67%, recycled.)

In recycled paper manufacturing, there is no fixed formula used to dictate the percentages of fiber supplied by wastepaper. Moreover, mills sometimes change the source of their fiber furnish on different runs of the same paper, depending on the economics or availability of various pulps. Some paper grades may be made at different times with 50% recycled fibers, 100% recycled fibers, or some other percentage altogether. (It is, of course, impossible to achieve a 100% recycled content by total paper weight since the fillers and additives are not recycled.)

The source of the wastepaper used— whether it is from internal mill wastes, purchased pulp substitutes, purchased industrial and printing plant wastes, or consumer sources— varies a great deal depending on the manufacturer and the particular type of paper being made. Some wastes may be deinked; others may not. There are no tests available to determine how much recycled content is present. Even use of a microscope will not reveal whether the cellulose fiber has been used once, twice, or more. Buyers who wish to specify paper on the basis of percentage recycled content and the types of wastepaper consumed must know how to ask appropriate questions of their suppliers. Currently, in the United States, in the absence of any nationwide verification programs, specifiers must depend on the accuracy and honesty of the suppliers who provide this information.

Deinking Technology and Methods

While pulp substitutes are easily used to make recycled paper, printed paper wastes present a much greater challenge. Customarily, though not always, this wastepaper is deinked if the fiber is to be used in printing and writing papers. A few papers are made using printed paper wastes that are not deinked but simply repulped and used directly. There are limitations, however, to the types of wastes that can be used in such a manufacturing process. On an industry-wide basis, deinking will always be required if significant quantities of high-grade wastepaper are to be employed in the production of printing and writing papers.

Deinking is the process of removing printed inks and applied finishing materials from the reusable cellulose fiber of the paper. It is a complex process. There is no one rigid formula for the design of deinking systems, and the specific types of wastepaper to be utilized have a great bearing on how any ideal system would be constructed. Most deinking systems in the printing and writing industry have been continually upgraded over the

80
Arthur C. Veverka, "Economics Favor Increased Use of Recycled Fiber in Most Furnishes," *Pulp & Paper,* Vol. 64, No. 9 (September 1990), pp. 97–103.

81
Contaminants that make it to the paper machine can plug the wires and felts of the machine or be melted onto the dryer drums. This sometimes causes dramatic breaks in the rapidly traveling paper rolls during manufacturing, or it may simply result in the formation of blotches and holes in the finished paper.

years, employing a range of technology and equipment that has improved over time and is rapidly changing today as greater resources are being put into new engineering developments. The installation of new deinking equipment is quite expensive, though recent studies have shown that the construction and operating costs for new deinked pulp mills can compare favorably with the costs for new virgin pulp mills.[80]

The deinking process has been used in papermaking for a long time, but it has been made more difficult in the past few decades by the proliferation of new printing and graphic arts processes— photocopying and laser printing technologies, the growing number of different types of offset, lithographic, and flexographic printing inks, ultraviolet and heatset coatings, pressure-sensitive adhesives, hot-melt glues, and so on. In addition to these "contaminants," which by themselves are enough to give any papermaker nightmares, there are many other real contaminants that inevitably show up in wastepaper bales. Styrofoam cups, plastic bags, and other synthetics are particularly problematic because, if they are not found and removed prior to pulping, particles of these materials are very difficult to remove in any screening and cleaning process.[81] In addition, mills report finding watches, clothing, and a variety of other strange items in the baled wastepaper they purchase.

Pulping
Deinking is accomplished by applying three forms of energy— mechanical, chemical, and thermal. This begins in the pulping stage. More than a simple blending of wastepaper and water, this step involves carefully establishing conditions that directly influence the effectiveness of the entire process. The high-consistency pulpers preferred today operate with the addition of less water than older equipment. The temperature of the stock is raised moderately; the heat, water, and added chemicals help swell the fibers. The attached inks and adhesives on the surface of the fiber then start to crack off. Friction is created by the fibers rubbing against each other as this thick stock is blended and churned, and this action further separates ink from fiber.

After pulping, a modern deinking system utilizes three principal processes for ink removal: washing, flotation, and dispersion. In addition, coarse and fine mechanical screens trap larger and heavier extraneous materials, and centrifugal or other specialty cleaners remove smaller, lightweight

Wastepaper riding up the conveyer belt for repulping.

Figure 3.6
Simplified Diagram for a Fine Paper Deinking System

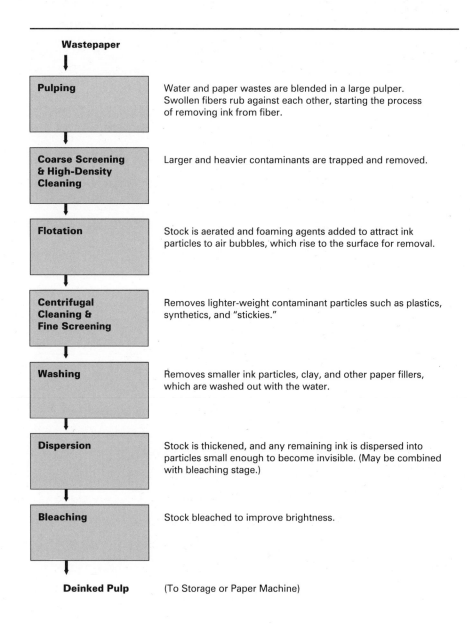

Wastepaper
↓

Pulping — Water and paper wastes are blended in a large pulper. Swollen fibers rub against each other, starting the process of removing ink from fiber.

Coarse Screening & High-Density Cleaning — Larger and heavier contaminants are trapped and removed.

Flotation — Stock is aerated and foaming agents added to attract ink particles to air bubbles, which rise to the surface for removal.

Centrifugal Cleaning & Fine Screening — Removes lighter-weight contaminant particles such as plastics, synthetics, and "stickies."

Washing — Removes smaller ink particles, clay, and other paper fillers, which are washed out with the water.

Dispersion — Stock is thickened, and any remaining ink is dispersed into particles small enough to become invisible. (May be combined with bleaching stage.)

Bleaching — Stock bleached to improve brightness.

Deinked Pulp (To Storage or Paper Machine)

contaminants. The precise sequence of steps varies a great deal from plant to plant, and systems are generally designed so they are not completely linear. Mill personnel may be able to reroute the stock through certain loops or through the entire process if necessary.[82]

Washing

Washing is the oldest technique used for removing ink; virtually all deinking mills have washing equipment. This method is sometimes compared to the washing of clothes. Detergents, wetting agents, and other chemicals are added to make the ink hydrophilic (attracted to water). The washing action helps to break up the ink particles, and repeated rinsing removes these particles with the wastewater, leaving the fiber behind. Washing systems effectively remove some of the smallest ink particles, (typically those under 30 microns). Washing systems alone were effective for deinking in earlier decades when inks were formulated more simply and there were fewer different inks, coatings, and adhesives. Today, washing systems are an essential component of a deinking plant, but by themselves they are insufficient for dealing with the wastepaper stream.

Flotation

Flotation deinking techniques were developed about 30 years ago in Europe. This technology was relatively little used in the United States until about 1975, but since then many deinking plants have added flotation equipment to their systems. Flotation is now considered an essential component of any new deinking plant.[83] In contrast to washing, flotation involves the addition of foaming agents and chemicals to make the ink hydrophobic (repelled by water). The stock is then continuously aerated, and the air bubbles attract the hydrophobic ink particles as they rise up to the surface. At the top, the foam of air bubbles carrying the ink particles is floated off or removed, leaving deinked pulp behind. Flotation systems are most effective at removing ink particles somewhat larger (about 30–100 microns) than those removed by washing.

Modern deinking plants employ both flotation and washing. Some position washing before flotation, and others do it in reverse. Because washing chemistry requires making the ink particles hydrophilic and flotation chemistry requires making them hydrophobic, plants using both systems must carefully balance the transition from one system to the other.

82
The summary of deinking technology and processes on these pages is based on research from a broad spectrum of sources. See the series of articles titled "Special Report: Deinking," *Pulp & Paper,* Vol. 64, No. 3 (March 1990), pp. 71–85, and the even longer series of articles titled "Engineering: Recycling," *Pulp & Paper,* Vol. 64, No. 9 (September 1990), pp. 75–230. See also James P. Casey, ed., *Pulp and Paper Chemistry and Chemical Technology,* Vol. 1 (3rd ed.), pp. 568–594. Additional information was provided by E.H. Pechan & Associates, "Survey of Deinking Research and Development in the North American Paper and Paperboard Industry," Springfield, VA, 1990. Personal visits were made to two different printing and writing mills operating deinking systems. Extensive telephone interviews were conducted with the managers or key personnel at all U.S. deinking operations currently supplying pulp to the printing and writing industry.

83
William P. Carroll and Michael A. McCool, "Pressurized Deinking Module," paper presented at *New England TAPPI/Connecticut Valley Pima Technical Seminar,* Holyoke, MA (March 1991).

Washing is an essential component of any deinking operation. These "sidehill washers" are part of the deinking plant at Cross Pointe's Miami Paper Mill, where many different uncoated printing and writing papers are made.

Flotation deinking is performed at the Miami Mill in these open flotation cells. Other flotation equipment is designed to perform this same function in a closed pressurized container. In both cases, air bubbles up through the cell, attracting ink particles in the watery pulp; the ink-containing foam that collects at the surface can then be readily separated from the pulp.

Centrifugal cleaners remove smaller lightweight contaminant particles, such as plastics and "stickies," that cannot be removed by mechanical screening processes.

84
See the section, *A Comparison of Virgin and Recycled Paper Manufacturing,* later in this chapter, for further discussion of bleaching issues. See also chapter 2 for a brief explanation of the different methods developed to pulp wood, as well as Appendix 2, *Pulping & Papermaking Processes.*

Dispersion

Dispersion units can be employed to contend with inks remaining after washing and flotation. New equipment has been developed in the last five or six years that operates on the principle of breaking up residual ink particles into a small enough size as to be invisible to the naked eye. This residual ink may dull the overall brightness of the sheet slightly, but no ink specks would be seen. Dispersion equipment is very energy-intensive, and thus expensive to operate, and not all mill operators are convinced of the overall efficiency or effectiveness of this treatment. This technology is not yet commonly employed in printing and writing mills with deinking systems. Because dispersion units operate at raised temperatures with a thickened pulp stock, they can be combined with the bleaching process.

Bleaching

Deinked stock is generally bleached to improve its overall brightness, further stripping residual inks and dyes left on the fiber. Deinked pulp requires substantially less bleaching than virgin chemical pulps, however. Different bleaching chemicals and processes are required depending on whether the pulps are being made with groundwood or chemical (groundwood-free) content.[84]

Deinking Mills

Few deinking grade wastes are used in printing and writing papers because, at present, there is very little industry capacity to utilize them. There are over 150 paper mills in the United States producing, collectively, over 22 million tons of printing and writing papers each year. As of mid-1991, however, there were only six individual mills producing printing and writing papers as their primary product that operated true deinking systems with the equipment necessary to use printed wastepaper for a significant portion of their fiber furnish. A seventh mill has a new plant under construction. There are also two mills that are primarily producers of newsprint from 100% deinked pulp but also manufacture some uncoated groundwood papers. The combined daily capacity of these plants to produce deinked pulp destined for use in printing and writing paper production is approximately 1,100 tons.[85] This represents less than 2% of the total daily pulp requirements for the production of printing and writing papers.

Two of these mills have been using recycled fiber extensively and deinking since the early part of this century. The Miami Paper Mill in West Carrollton, Ohio, one of two mills now owned by the Cross Pointe Paper Corporation, has a modern deinking system with both flotation and washing technology. The mill supplies one-half of its entire fiber needs from this deinking plant, which has been in operation since about 1915. Cross Pointe is continually making investments to upgrade its deinking plant and has been well known for some time as a producer of high-quality recycled papers made with a substantial percentage of deinked wastes. In Neenah, Wisconsin, the Bergstrom Paper Division of the P.H. Glatfelter Company operates a paper mill with the largest daily deinking capacity of any printing and writing mill in the country. This mill is also able to produce more than 50% of its total fiber requirements by deinking. Both Cross Pointe and Glatfelter make a variety of uncoated and film-coated offset papers for book publishing, commercial printing, converting, and other uses.

The Georgia Pacific Corporation, one of the largest U.S. paper and forest products companies, has many mill operations making a wide variety of paper products, including about ten mills that make printing and writing papers. One of these, in Kalamazoo, Michigan, has a deinking facility.

[85] Deinking capacity data are from *Pulp & Paper North American Factbook: 1990,* and verified by direct or telephone interviews with key personnel at each of the deinking mills.

This mill has been deinking since 1960 and was one of the first printing and writing mills to install a flotation deinking system. The Kalamazoo mill distributes deinked recycled fiber throughout virtually all grades of paper it makes, though the company only classifies some of its paper lines as recycled. The Simpson Paper Company has nine paper mill operations in the United States. One plant, the San Gabriel Mill in Pomona, California, has a deinking system that has been in operation for about 20 years. While the daily capacity of this deinking system is somewhat smaller than those previously mentioned, Simpson does produce a portion of the deinked fiber they use for some of their recycled grades.

Table 3.7
North American Printing and Writing Mills Operating Deinking Systems

Mill	First Date of Deinking Operations	Deinked Pulp Capacity	Paper Production Capacity	Recycled Paper Products from These Mills
Appleton Papers West Carrollton Mill West Carrollton, Ohio	circa 1950	180 tons/day	350 tons/day	Carbonless paper
Cross Pointe Paper Miami Paper Mill West Carrollton, Ohio	circa 1915	160 tons/day	300 tons/day	Uncoated offset, book publishing, text & cover, bond, and writing papers
Georgia Pacific Kalamazoo Mill Kalamazoo, Michigan	1960	140–160 tons/day	300–380 tons/day	Uncoated offset and bond papers
Noranda Forest Recycled Papers Thorold, Ontario Canada	1954	110 tons/day	200–250 tons/day	Uncoated offset, coated cover, reply card, writing papers
P.H. Glatfelter Co. Bergstrom Paper Division Neenah, Wisconsin	circa early 1900	250 tons/day	410–450 tons/day	Uncoated offset, book publishing, text, bond, and converting papers
Patriot Paper Hyde Park Mill Boston, Massachusetts	1960s under former owners	200 tons/day	200 tons/day	Uncoated offset, text & cover, bond, copier, and index papers
Riverside Paper EcoFibre Division DePere, Wisconsin	Under construction, operational in 1992	80–100 tons/day	120–150 tons/day at Appleton Mill	Uncoated offset, tablet, writing, and bond
Simpson Paper San Gabriel Mill Pomona, California	1970	60–65 tons/day	340 tons/day	Uncoated offset, text & cover, copier, and bond papers

Two producers of recycled newsprint from 100% deinked pulp also make a limited number of grades of uncoated groundwood papers: **Manistique Papers** (Manistique, Michigan) and **FSC Paper Company** (Alsip, Illinois).

In Boston, Massachusetts, the newly formed Patriot Paper Corporation began papermaking operations in the fall of 1990, after purchasing an idled paper mill with older deinking equipment that had not been used for seven years. Patriot is bringing this system back on-line while simultaneously investing $27 million in new deinking equipment. When complete, this system is expected to be one of the most sophisticated deinking plants in operation at any of the mills making printing and writing papers in this country.

Appleton Papers, one of the largest producers of carbonless paper in the world, operates a deinking facility at its mill in West Carrollton, Ohio, one of its three mills making these specialty papers. The Riverside Paper Company in Wisconsin has a deinked pulp mill under construction in DePere that is expected to come on-line in 1992; the cost of this plant will be on the order of $17.5 million.[86] Riverside currently utilizes wastepaper in a limited fashion, but the construction of this new plant will give this company the full capacity to produce all of the deinked pulp required for its own production of recycled papers.

The only other printing and writing mill in North America with its own deinking plant is operated by Noranda Forest Recycled Papers in Ontario, Canada. This system has been operational for about forty years, and the company has planned a major expansion to be undertaken when economic conditions permit. Currently the mill produces enough deinked pulp to manufacture a range of book publishing, offset, and converting papers in which over 50% of the total fiber content comes from deinked wastepaper.

Paper Mills without Deinking Systems

Paper mills without their own deinking capacity are primarily dependent upon the purchase of deinked market pulp from an outside supplier in order to make recycled paper with deinked or postconsumer content.[87] The supply of deinked market pulp is relatively limited, and mills that purchase it tend to use a much smaller percentage of deinked pulp in their papers than do paper mills that operate their own deinking systems. Until late 1990, essentially the only U.S. supplier selling deinked pulp to printing and writing mills was Ponderosa Fibres of America.[88] This corporation has four deinking plants, in Santa Ana, California; Memphis, Tennessee;

---

[86] "News in Brief," *Paper Recycler,* Vol. 2, No. 8 (August 1991), p. 11.

[87] See chapter 4 for a complete discussion of postconsumer waste issues.

[88] Ohio Pulp Mills in Cincinnati, Ohio, also makes market pulp from wastepaper, at a capacity of only 35 tons per day. This facility "de-polys" stock such as polyethylene-coated milk cartons but does not have a full deinking facility. In Canada, the Desencrage Cascades Mill in Quebec makes 100 tons of deinked market pulp per day. See "Manufacturers Expand Recycling Programs for Milk, Juice Cartons," *Paper Recycler,* Vol. 1, No. 1 (October 1990), p. 9, and "Deinked Market Pulp Provides Opportunities for Buyers and Sellers," *Paper Recycler,* Vol. 1, No. 2 (November 1990), pp. 5–6.

Oshkosh, Wisconsin; and Augusta, Georgia. Ponderosa Fibres commenced operations over 25 years ago for the express purpose of supplying deinked pulp to the tissue industry. The transition to supplying fiber for printing and writing mills has been comparatively recent; it was only a tiny portion of their business until the late 1980s. Since the growth of consumer demand for deinked content in printing and writing papers, there has been a dramatic shift in Ponderosa's business toward these customers. By early 1991 Ponderosa was selling 50–70% of its combined daily output of 700 tons of deinked pulp to printing and writing mills, and this company remains the principal supplier of market deinked pulp to the printing and writing industry.[89]

There are a few other limited sources of supply to which printing and writing manufacturers without their own deinking systems may now occasionally turn. Again, this has come about as companies have felt a need to add such pulp to their papers in order to market them not only as recycled but as including a "postconsumer" content. Occasionally printing and writing mills purchase deinked pulp from existing tissue or newsprint mills with wastepaper deinking plants. In addition, in late 1990, the Mississippi River Corporation started producing deinked market pulp at a formerly idled mill in Natchez, Mississippi, and its customers include mills producing printing and writing papers.[90] Several new deinked market pulp operations have also been announced as being underway, planned, or under consideration. Future expansions in the use of deinked pulp in the printing and writing paper industry are as likely to utilize new independent suppliers of market pulp as they are to involve the installation of major new deinking plants at existing mills.

The advantage to operators of these deinked market pulp mills is that they can sell their product to a wide range of paper producers. If a particular run of pulp does not meet the higher standards required for printing and writing papers, it may be suitable for a tissue mill or other buyer. Thus market pulp mills have some flexibility, a helpful advantage given the difficulties inherent in using wastepaper as a fiber source. The advantage to a printing and writing mill that buys market pulp from an outside supplier is that it has not had to make the huge investments necessary for the construction of a deinking plant. On the other hand, without direct control over the quality of the pulp purchased, the mill may find it more difficult to achieve the quality necessary in the production of high-grade papers.

89
Kelley H. Ferguson, "Ponderosa Fibres Expands to Meet Demands for Recycled Market Pulp," *Pulp & Paper,* Vol. 64, No. 9 (September 1990), pp. 212–214, and telephone interview with Glen A. Tracy, Ponderosa Fibres Assistant Vice President, March 1991.

90
"Deinked Market Pulp Provides Opportunities for Buyers and Sellers," *Paper Recycler,* Vol. 1, No. 2 (November 1990), pp. 5–6.

Direct-Entry Deinking

A few mills, together with chemical suppliers to the paper industry, have been working on methods for deinking wastepaper directly in the pulper by chemical processes, avoiding the need for the extensive washing and flotation equipment normally associated with a deinking plant. Again, the reason for such an approach is to allow the inclusion of deinked or postconsumer content in the paper production without making the large capital commitments required to install a deinking plant. The objective of direct-entry deinking is not so much to remove ink as it is to reduce the ink particles to a small enough size through chemical means, so that they are dispersed throughout the pulp and invisible to the naked eye. This approach poses major challenges to manufacturers because the chemistry required for direct-entry deinking has the potential to adversely affect the papermaking process itself, and sometimes even the printability of paper that is produced. There are also real limitations on the total amount and types of wastepaper that can be used, and on how heavily printed the wastepaper can be on average. This process, however, has potential for success in very specific and carefully controlled situations.[91]

Deinking Sludges

The process of deinking leaves a residual sludge after the usable cellulose fiber has been extracted. This sludge consists of water, fillers, inks, short fibers, and other nonrecoverable portions of the wastepaper. Questions have been raised about the environmental effects of disposing of these sludges. These questions seem to assume that if the existing printed wastepaper we generate is not deinked, all is well. Perhaps the inks, and other printed and applied contaminants, somehow magically disappear, thus ceasing to have an effect on our environment. This is far from the truth.

There are three disposal options for printed paper products not kept as part of permanent records: landfilling, incineration, and recycling. In a landfill, the inks have the potential to leach into and contaminate the groundwater. In an incinerator, combusted inks generate both airborne pollutants and residual contaminants in the remaining incinerator ash, which is itself subsequently landfilled. In contrast to these first two options, several benefits result from the deinking process. First, the usable and most substantial portion of the paper, the cellulose fiber, is extracted for recycling. Second, the cost and responsibility for disposing of the remaining

[91] Roger J. Dexter, Betz PaperChem Recycled Fiber Group, "Direct Entry Deinking," paper presented at *New England TAPPI/Connecticut Valley Pima Technical Seminar*, Holyoke, MA (March 1991).

wastes, the sludge, is now borne by industry rather than public municipalities. One could argue that, by manufacturing paper through deinking, industry is in fact performing a public service for the cities and towns that otherwise are left to dispose of these wastes.

Each mill must have its deinking sludges tested for toxicity. Upon testing, most are classified as nonhazardous. Many mills dispose of their sludge in lined landfills with leachate collection and groundwater monitoring systems. Most states now require this type of construction as part of their permitting process. Such disposal standards are often stricter than those met by operating municipal landfills. Other options for disposing of sludge include burning, reuse (in concrete manufacture for example), or "landfarming" (distributing the sludge over farmland soils). Probably none of these methods is completely without risk. The concern with landfarming is that residual heavy metals or other contaminants from the inks might accumulate in the soil.[92] The bottom line is that the formulation of inks and other materials needs to be done in as benign a manner as possible, for once they are manufactured, they will have some impact. The ink industry has, in fact, substantially reduced the use of heavy metals in inks over the past few decades, and the total amount of these compounds in inks now tends to be relatively small. Presumably, improvements will continue to be made.[93]

The Office Wastepaper Problem

Photocopies and laser-printed materials pose unique problems to deinking mills. Because of the similarity of the technologies, these two products are known collectively by the paper industry as "laser inks" or simply "laser." Unlike conventional printing inks, which consist of pigments suspended in liquid solution and subsequently applied to paper directly or by an offset process, laser imaging represents a completely different technology. Both photocopiers and laser printers employ dry powder toners consisting largely of plastic polymers, which are transferred to the paper by an electrostatic charge and then require heat fixing to be fused in place. The resulting image essentially consists of plastic particles that have been melted together and simultaneously bonded to the paper fiber. These products are thus resistant to deinking. If the wastepaper used by a deinking plant contains laser print, ink specks will almost inevitably make their way into the finished paper. Some mill executives have reported seeing whole letters

[92] Robert C. Carroll and Thomas P. Gajda, "Mills Considering New Deinking Line Must Answer Environmental Questions," *Pulp & Paper,* Vol. 64, No. 9 (September 1990), pp. 201–205.

[93] See the section on *Ink Coverage and Color Choices* in chapter 6 for additional discussion of heavy metals in inks.

[94] Richard Strauss and Fred Iannazzi, "Problems and Opportunities in the Utilization of Recycled Printing & Writing Paper," paper presented at *New England TAPPI/Connecticut Valley Pima Technical Seminar*, Holyoke, MA (March 1991).

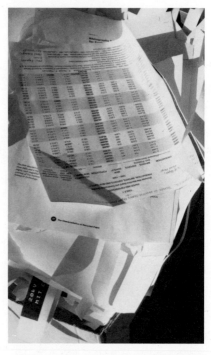

Office wastes represent one of the largest sources of recycled fiber potentially available to the paper industry. Currently, few of these wastes are used.

that remained fused together show up in their finished paper! While this may not be so terrible for tissue grades, most printing and writing mill executives are not enthusiastic about selling fine paper with these imperfections.

For this reason, operators of deinking plants at printing and writing mills typically do everything they can to avoid these wastes. Wastepaper buyers at these plants specify "laser-free" wastes and reject bales that arrive with any significant quantity of laser print inside. A great deal of research is currently being done on ways to manage these contaminants. Both Betz PaperChem and PPG Industries are developing chemistries that cause laser inks to agglomerate into particles large enough to be removed and screened out of the pulp by the cleaning equipment within the deinking system.

As newer and more sophisticated deinking plants are built, and as existing deinking plants are upgraded, it will become more possible to use office wastes containing laser print in the furnish. Hypothetically, the technology currently exists to construct a plant that could be successful at deinking these mixed wastes and creating a pulp perfectly free of specks or dirt count. From a practical standpoint, because such a system would require an extensive array of equipment and the latest technologies, it would be quite expensive to operate. Most existing deinking systems, if they are to use any significant amount of postconsumer office wastes, require some customer tolerance for a visible dirt count. The modernization of the Patriot Paper Mill in Boston, Massachusetts, and the installation of new equipment at the Miami Mill of the Cross Pointe Corporation, are expected to bring systems on-line that will be as good as any at utilizing office waste, though the final capabilities of these plants are not yet known.

Meanwhile, the wastepaper discarded in offices across the United States represents about 40% of all printing and writing paper sold. Presently, very few of these wastes (just over 10%) are being recovered, leaving 8.4 million tons of office wastepaper discarded into the MSW stream each year. Even at these low collection rates, managers of office recycling programs sometimes have a hard time finding dealers to take their collected material. Thus, office wastepaper, despite being one of the largest segments of the paper waste stream, remains a relatively untapped resource that could yet be a major source of fiber for the paper industry.[94]

An interesting approach to the office wastepaper problem has been developed by several mills that are carefully selecting certain grades of office waste and pulping it directly without deinking. In this process, the printed ink and contaminants on the wastepaper are ground up to become a textural part of the new paper being made. Generally this results in a paper with a scattering of small, dark ink specks throughout, and the result can be aesthetically pleasing. If the paper wastes used are lightly enough printed on average, and if the mills are able to maintain consistency in their wastepaper sources, production of high-quality paper can be achieved.[95] While there are limitations to the types of wastes used, this approach is a desirable utilization of resources in a fashion that minimizes environmental impact. It is an appropriate way to create a paper in which texture is desired.

That Recycled Look: Fiber-Added Papers

It is a misconception that recycled papers are always full of flecks and specks. *Printed paper wastes, if properly deinked, can provide a high-quality pulp to make fine bright papers, free of dirt specks and imperfections.* Many of these papers have been sold for years in the same market as virgin papers, without being differentiated as a recycled product. Customers have been unaware of any difference in fiber sources. Yet flecked papers have come to be associated with the recycled paper industry. Ironically, almost all textured papers are created by the addition of distinctive colored fiber to the paper for aesthetic reasons that have nothing to do with the recycling process at all. The few textured papers made from non-deinked printed wastepaper are the exception.

There are many papers made, both virgin and recycled, that have fibrous material added to give distinctive texture and color to a particular grade. These are referred to as "fiber-added" papers. If the flecks are brown, slightly irregular, and varied in size, they are most likely wood shives, which may be purchased from virgin wood pulping operations. The shives are part of the rejected fiber collected after pulping is completed. If the flecks are hairlike threads, often brightly colored, they are likely fine cotton threads purchased from the textile industry. Papers made with such added threads are also referred to as "flocked." Fiber-added papers are even made with planchettes— small punched-out pieces of colored paper, usually circular— added to give a festive confetti-like feeling or, in the case

---

95
Domtar Papers in Canada has been noted for a grade of paper made in this manner for several years. Simpson Paper Company in the United States has recently introduced a paper made from 100% non-deinked wastepaper including office wastes with laser-printed materials. Several other mills in the United States are using non-deinked wastes for some portion of their fiber furnish in limited grades of paper.

of currency papers, to protect against counterfeiting. Generally, these planchettes are punched by the paper mill specifically for this purpose, so that they can control the desired aesthetic parameters including color, size, and frequency. Contrary to what one might logically assume, they are not the leftover "holes" punched during the manufacture of computer forms or other paper punching and drilling operations.

When recycled papers are made with printed wastepaper and the deinking operation has been less than completely successful, the finished paper may show small dark ink specks scattered in the sheet. Unlike the case of fiber-added papers, this is not intentional. These papers have a visible dirt count.[96] This problem is more likely to occur if the deinking plant lacks flotation equipment, if heavily printed wastepaper is used, and especially if laser print is in the furnish. If this pulp is subsequently used to make fiber-added paper, the dirt specks may be masked by the texture of the added fiber. Thus papermakers might choose to make a recycled fiber-added sheet both for aesthetic reasons and for the cloaking advantage this can provide.

Cotton Fiber Papers

Papers made with 25–100% cotton fiber are classified as cotton papers. Prior to 1900, cotton supplied the majority of the fiber used in the United States for papermaking. Today, given its limited supply and high cost compared to wood, cotton supplies less than 1% of all fiber used for domestic production of printing and writing papers.[97]

Cotton is one of the finest materials for papermaking. It is used for many specialty papers and for papers requiring exceptional permanence and durability, especially currency papers, certificates, historical documents, and some other business papers.[98] Unlike wood, which has a high percentage of lignin and is roughly only one-half cellulose, the seed hairs of the cotton plant are 92% cellulose, and they contain virtually no lignin.[99] Because of its high cellulose content and naturally light color, cotton processed for pulp requires substantially less chemical treatment and bleaching than does wood. Historically, 100% cotton papers were made for a long time without any bleaching, and they could still be made this way today if both mills and buyers were willing. The resulting color would be a pleasant, creamy shade of off-white or light beige, somewhat similar to unbleached muslin.

96
Quality standards in paper manufacturing, whether virgin or recycled, include a dirt count "test," which consists of simply counting any visible foreign specks or contaminants over a standardized area. Each mill may have its own set of standards for maximum acceptable dirt count.

97
American Paper Institute, *Paper, Paperboard, and Wood Pulp Capacity: 1990*, p. 8.

98
U.S. currency papers are made with a combination of cotton and flax pulp. Cotton is the primary source of fiber, but about 15% flax pulp is included for greater durability.

99
Robert R. A. Higham, *A Handbook of Papermaking* (London: Oxford University Press, 1963), p. 9.

Ripe cotton bolls contain many seeds (making up about two-thirds of their bulk) and long, white fibers. During ginning the seeds are separated from the fibers so that the fibers may be spun into thread. The seeds are not discarded; they are separately processed to remove the linters, the short fuzzy fibers that cover them, and to extract cottonseed oil. Cotton linters have many commercial uses, of which papermaking is an important one. Cotton linters supply the majority of cotton fiber now used in the manufacture of cotton papers. In addition, some cotton fiber in papermaking comes from other sources. For example, the Crane Paper Mill uses discarded fiber collected from the cotton ginning operation, as well as cotton textile and garment clippings. In late 1991 Crane even began using some discarded cotton products (e.g. sheets and tablecloths) from postconsumer sources.

100

Casey, ed., *Pulp and Paper Chemistry and Chemical Technology*, Vol. 1 (3rd ed.), pp. 594–601.

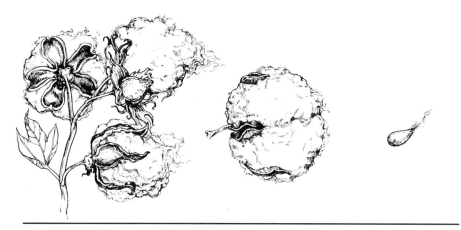

Ripe cotton bolls at harvest time.

Burst cotton boll, containing fiber, lint, and numerous seeds.

Single cotton seed covered with short, fuzzy fibers, known as "linters."

Almost without exception, cotton papers are no longer made of old rags. The term "rag bond" is a carry-over from days when old rags were the source of fiber. The cotton fiber used in papermaking today comes from textile mill and manufacturing cuttings, leftover seed hairs culled from the cotton ball during the ginning of cotton into thread, and cotton linters. Linters are the short fuzzy fibers that remain firmly attached to the cotton seed after ginning. Salvaging the linter fibers requires physically cutting the seeds, and this is usually done by companies that are processing the seeds for cottonseed oil. Today, because of the ubiquitous synthetic threads and other finishes used by the textile and clothing industries, textile cuttings provide a relatively small portion of the fiber used for cotton pulps, while cotton linters provide over 50% of the fiber used by this segment of the paper industry.[100]

There has been some debate about whether cotton fiber should qualify as recycled content. Certainly, it can be argued that textile garment cuttings, as trimmings from an industrial manufacturing operation, are comparable to paper trimmings generated by printers, binderies, and finishing houses creating printed products. Cotton linters and scrap cotton fiber collected during the ginning operation are also by-products of one industry subsequently purchased for use by another. This debate aside, it is clear that because cotton is so well-suited for papermaking, there are environmental benefits inherent in the use of this plant material as a fiber source.

A Comparison of Virgin and Recycled Paper Manufacturing

Resource Use and Tree Consumption

It has been widely reported that every ton of paper, if made from 100% recycled fibers, saves approximately 17 trees.[101] In reality, the number of trees required to generate a ton of paper depends on many factors including tree size, species, and the type of pulp and paper being manufactured. Groundwood papers, because they do not have the lignin and other wood components removed, require less wood per ton than do free-sheet papers. Free-sheet papers made from chemical pulps, such as offset printing, bond, and most other printing and writing papers, require more wood per ton of paper production than do groundwood papers. Free-sheet papers, however, are more suitable for recycling into a wider variety of new papers because of the relative purity of the cellulose fiber in the original pulp.[102]

Despite the difficulty of approximating the number of trees required per ton of paper production, we humans need a measure of some sort with which to understand the implications of our actions. A recent estimate yields a range of 17–31 trees per ton, based on the following assumptions, which are intended to be representative of printing and writing paper production. The trees used were a mixture of softwoods and hardwoods 40 feet tall and 6–8" in diameter (a size at which pulpwood trees are often harvested). The pulping yield was based on an average process loss more characteristic of chemical pulping processes than semi-chemical or groundwood processes.[103] Very roughly, it may take an average of 24 such trees to produce a ton of printing and writing paper. The widely used figure of 17 trees is thus conservative for printing and writing papers. Yet, regardless of the exact number of trees felled for each ton of paper, the simple fact remains that recycling used cellulose fiber will save a significant amount of wood resources. In light of the fact that the United States is currently consuming about 100 million cords of wood annually for pulpwood and that this consumption is still growing, the need to save wood resources is apparent.

Subsidies That Support the Virgin Timber Industry

There are several ways in which the existing U.S. tax structure creates a competitive advantage for the manufacture of virgin paper. First, 22% of the country's wood harvest, some of which is destined for paper production, comes from public lands, with the majority of these lands administered by the U.S. Forest Service. A detailed independent analysis of timber sales from the national forests shows that actual revenues from

---

[101] *Get Real! A Consumer Guide to Real Recycled Paper* (San Francisco: Conservatree Paper Company, 1990). The origin of this statistic and the size of the trees used to generate this estimate are unknown.

[102] Groundwood-containing papers are prone to more rapid deterioration than free-sheet papers, and they are suitable primarily for recycling into new newsprint or new groundwood papers. Papers made from chemical pulps have the potential to be readily recycled into almost any type of paper product.

[103] This estimate was researched and developed by Tom Soder, Pulp and Paper Technology Program, University of Maine, September 1991.

sales are typically not sufficient to cover the government's expenditures for road building, reforestation, and other sale-related activities. According to The Wilderness Society, during FY 1990, U.S. taxpayers effectively subsidized the timber harvest from 98 of the 120 national forests in which sales were conducted, at a cost to the federal treasury of $257 million.[104]

Other indirect tax benefits that support the use of virgin wood include the capital gains treatment of timber sales, and a depletion allowance based on treating the cutting of timber as the depletion of a nonrenewable instead of a renewable resource. Limited deductions for reforestation expenditures are also allowed.[105] Recently, there have been some limited tax incentives to support the development of recycling, principally in the form of favorable municipal financing for the construction of new deinking plants, but the net effect of our economic structure is still weighted toward encouraging the exploitation of virgin materials rather than the recycling of existing materials.

Solid Waste Issues

Regardless of whether recovered paper wastes would have been landfilled or incinerated, recycling them instead is beneficial on several fronts. First, the communities or corporations that would have to pay for disposal save the tipping fees; these have come to be known as *avoided costs*. Additionally, as markets for wastepaper become more developed, the selling of these wastes, especially the large quantities of high-grade papers, may even generate income. From a disposal standpoint, since a ton of wastepaper occupies 3 cubic yards of landfill space,[106] every ton of paper specified, if made from 100% recycled fibers, saves about this much landfill space. (It actually requires more than a ton of wastepaper to produce a ton of deinked pulp, due to the process loss during deinking. Thus, even though some sludge is generated by the deinking mill, the net savings of landfill space remains roughly 3 cubic yards.) If the wastepaper recovered had been slated for incineration instead of landfilling, its use in recycling may still save the landfill space that would have been required for the disposal of the incinerator ash.

Energy Consumption

In absolute terms, the production of recycled paper consumes significantly less energy than the production of paper from virgin wood. One ton of pulp made from deinked and bleached wastepaper requires 60% less energy, or 14 million fewer Btus, to manufacture than does a ton of bleached virgin kraft pulp. This represents a substantial savings, equivalent

---

104
Richard Rice, *Taxpayer Losses from National Forest Timber Sales, FY 1990*, Wilderness Society Forest Policy Update (Washington, DC: The Wilderness Society, May 1991).

105
Wendy Gerlach, Ropes & Gray, Memorandum outlining principal tax benefits available to the timber industry, June 4, 1990. Prior to the Tax Reform Act of 1986, the capital gains treatment of timber sales was the most significant benefit available to the virgin timber industry. Currently this benefit is less important due to the fact that capital gains and ordinary income are generally taxed at the same rate. Any future cuts in the capital gains tax would once again change this situation, however.

106
American Paper Institute, *Paper Recycling and Its Role in Solid Waste Management*, p. 4.

to 4,100 kilowatt hours of electricity, or approximately three-quarters of the amount consumed by an average residential household over an entire year for lighting and household appliances (not including heating and cooling units).[107] Virgin paper manufacturers will argue that such an energy comparison is unfair, though, because wood pulping by-products are now used to generate a large portion of the energy consumed at virgin pulp mills. Overall, the paper industry relied on purchased energy for only 42% of its power supply in 1989.[108] As a result, if one compares only purchased energy, the differences in energy consumption between virgin and deinked pulp mills supplying the printing and writing industry are not nearly as great. However, since the process of generating energy itself includes environmental consequences such as the release of CO_2 and other greenhouse gases, comparisons of total energy consumption are still important.

Air Pollution

A great deal less air pollution is generated in the production of recycled papers. In 1974 the EPA reported that for every ton of deinked and bleached recycled pulp produced, 58 fewer pounds of air pollution effluents were released into the atmosphere than would have been released for an equivalent ton of virgin bleached kraft pulp.[109] The majority of all virgin printing and writing papers are produced through kraft pulping methods, and it is this process that produces the sulfurous "rotten-egg" smell associated with papermaking. Modern pollution abatement equipment, however, can control much of these emissions.

Bleaching

Before bleaching, virgin wood pulp is the dark brown color we associate with "kraft" paper. In contrast, deinked pulp made from wastepaper is much lighter in color. The brightness of the deinked pulp will vary depending on the quality of wastepaper used, but pulp from the high-grade wastes typically employed to make recycled printing and writing papers will generally be a fairly light off-white or very light gray, with a brightness value around 60.[110] Thus, deinked pulps require substantially less bleaching than do virgin wood pulps, sometimes using as little as 10% of the chemical compounds required by virgin pulps.[111]

Bleaching has been a controversial issue of late, as concern has developed over the release of chlorinated organic compounds, and especially dioxins, associated with chlorine bleaching of paper. Some sources have falsely stated that chlorine bleaching is required for the manufacture of recycled

---

107

Energy comparison data for virgin and deinked pulp are from U.S. Environmental Protection Agency, Office of Solid Waste Management, *Resource Recovery and Source Reduction: First Report to Congress* (Washington, DC: Government Printing Office, February 1974), pp. 6–8.

Individual energy consumption can vary greatly, but an average U.S. household may consume 400–500 kwh of electricity per month for electric lighting and household appliances, not including electricity for heat, hot water, or air conditioning (interviews with energy staff at Com-Electric Company and data from Edison Electric Institute).

108

Pulp & Paper North American Factbook: 1990, p. 65.

109

U.S. EPA, *Resource Recovery and Source Reduction: First Report to Congress*, p. 8. This analysis included a review of transportation, manufacturing, and harvesting requirements.

110

Casey, ed., *Pulp and Paper Chemistry and Chemical Technology*, Vol. 1, p. 592. For a complete explanation of paper brightness values, see the section *Brightness Standards* in chapter 5.

111

Gordon Sisler, Manager of Product Development, Noranda Forest Recycled Papers, "Opportunities for Fine Paper Producers," paper presented at *Pulp & Paper Wastepaper I Conference*, Chicago, IL (May 1990).

papers, and that only virgin papers can be made "chlorine-free." In fact, a variety of bleaching processes can be used for either papermaking process. At present, the majority of all chemical pulps, both virgin and recycled, are bleached by processes that involve some chlorine-containing chemicals. It is possible, though presently rarer, for either virgin or recycled papers to be bleached without chlorine-containing compounds. Research is currently being directed toward developing and improving methods using oxygen, ozone, and hydrogen peroxide compounds. These methods generally achieve slightly less brightness in the resulting pulp.[112]

Most of the deinking mills supplying fiber for the printing and writing industry use wastepaper originally made with chemical pulp. Since the bleaching of this recycled pulp requires much less chemical use overall, there are significant environmental benefits associated with the bleaching of recycled rather than virgin fiber. In addition, the method used to bleach recycled fiber is typically a single-stage sodium hypochlorite process conducted in a high-alkaline environment. In this situation, dioxins are not formed.[113] This bleaching process is far safer than the multistage chlorine bleaching using chlorine gas and/or chlorine dioxide that is typically employed by virgin sulfate pulp mills, a process that is associated with dioxin formation.[114]

Jobs

It is generally acknowledged that collecting and processing recycled materials is more labor-intensive than is manufacturing using virgin materials. The harvesting of wood relies on the use of large, heavy equipment operated by relatively few people. Modern "super-integrated" paper mills are so streamlined and automated that once the logs are put in the system, they are chipped, cooked into pulp, bleached, and made into paper in one continuous process with essentially no handling from one end of the mill to the other. The manufacture of recycled paper, on the other hand, requires more human resources for the collection, sorting, and transport of wastepaper. Paper mills that are currently installing deinking operations are hiring additional personnel for this work. It is likely that increased production of recycled papers may create even more jobs.

112
Casey, ed., *Pulp and Paper Chemistry and Chemical Technology,* Vol. 1 (3rd ed.), pp. 592–594; interview with Kimmo Järvinen, Process Engineer, Kemira Peroxygen and Pulp Chemicals, March 1991.

113
E.H. Pechan & Associates, "Survey of Deinking Research and Development," pp. 19–20.

114
U.S. Congress Office of Technology Assessment, *Technologies for Reducing Dioxin in the Manufacture of Bleached Wood Pulp* (Washington, DC: U.S. Government Printing Office, May 1989).

4 Definitions and Standards

In the United States there are no national, legally binding definitions or standards governing the wastepaper content of recycled paper or the associated use of the term "recycled." Nor are there verification programs which consistently review the advertising claims of all manufacturers that market their papers as recycled. Consumers who wish to use recycled content as a criterion for purchasing are primarily dependent on the honesty and accuracy of the paper mills and distributors, as well as on their own ability to ask the detailed questions necessary in order to gain any real basis for comparison. There is an EPA Guideline, but it only governs purchases made with federal tax dollars. In the absence of any other nationwide standard, however, many mills have introduced grades of paper using this guideline as a minimum standard for recycled content. Many people view the EPA Guideline, which allows some internally generated mill wastes to count toward recycled content, as an ineffective model for a national standard. Thus, many state and regional governments have adopted their own, widely differing purchasing requirements, and a number of private organizations are promoting their own standards. This situation has created chaos for both the mills and purchasers. The Canadian government, in contrast, has instituted a clear and enforceable program governing the labeling and marketing of recycled papers sold throughout Canada, under the Environmental Choice EcoLogo program.

(Photo © Susan Larocque)

The EPA Guideline

In 1976 Congress passed the Resource Conservation and Recovery Act, now commonly called RCRA. Section 6002 of this law requires federal agencies, as well as state and local agencies using federal funds, to purchase designated products made with recovered materials if their annual expenditures for these items are more than $10,000. The law further directed the

U.S. Environmental Protection Agency to designate which products would be covered by this requirement and to develop guidelines to assist agencies in complying with the statute.[115] It was not until twelve years later, in June 1988, that the EPA finally issued its Guideline for Paper and Paper Products. This guideline required agencies subject to RCRA to be in compliance with these regulations within one year, by June 22, 1989. Specifically, these agencies are now required to take positive action to purchase recycled paper in preference to virgin paper, and to establish procedures to verify that their suppliers are meeting the minimum requirements for recycled content established for these federal purchases.[116]

The EPA Guideline covers almost all paper and paper products, and it establishes different minimum-content standards for the different types of paper products. For most types of printing and writing papers, the EPA recommends a minimum content of 50% wastepaper, and their definition of wastepaper is very broad. It includes *postconsumer* materials such as paper wastes collected from offices, retail stores, and residences, "after they have passed through their end usage as a consumer item."[117] It also includes a very large category of *preconsumer* wastes— ranging from paper mill manufacturing scrap to wastepaper generated from trimmings and rejects at printing plants, graphic arts finishing houses, envelope converters, and other manufacturers. As originally written, the EPA Guideline allowed any waste generated after paper has been rolled onto the winder at the end of the paper machine to count toward recycled content. The internal mill wastes that can thus be used for recycled content include the scrap generated in the trimming of these huge paper machine rolls into smaller web rolls and cut sheets, as well as obsolete inventories of paper, damaged rolls, and trimmings from other mill converting operations. These wastes have always been reused in the papermaking process, but the language of the Guideline led to a shift in the terminology used to describe them. Historically, such wastes were called "dry mill broke."[118] When the EPA defined mill broke more narrowly— limiting it to wastes generated "before completion of the papermaking process," many mill representatives simply stopped calling their paper converting wastes "mill broke" so that they could tell their customers that they do not use mill broke to meet recycled content requirements. But the fact is that many mills, though not all, commonly do use the dry paper wastes allowed under the EPA Guideline to provide a substantial portion of their "recycled" fiber.

115
To date the EPA has designated five products to which the statute applies: paper, lubricating oils, retread tires, building insulation, and fly ash in cement and concrete. Different guidelines cover purchases for each of these products.

116
Environmental Protection Agency, *Guideline for Federal Procurement of Paper and Paper Products Containing Recovered Materials; Final Rule, 40 CFR Part 250* (June 22, 1988), pp. 1–22.

117
Ibid., p. 4.

118
The Dictionary of Paper, published in 1980 by the American Paper Institute, defines *mill broke* as follows: "Paper that has been discarded anywhere in the process of manufacture. 'Wet broke' is paper taken off the wet press of a paper machine; 'dry broke' is made when paper is spoiled in going over the driers or through the calenders, trimmed off in the rewinding of rolls, trimmed from sheets being prepared for shipping, or discarded for manufacturing defects. It is usually returned to a repulping unit for reprocessing."

When originally written, the EPA Guideline did not specify whether the minimum wastepaper content was to be measured as a percentage of the total fiber content in the paper or as a percentage of the total weight of the paper. Since the recycled content percentages are higher when based on total fiber content, it quickly became common practice for mills to market their paper with wastepaper percentages based on this measurement rather than on total paper weight. The EPA has since clarified that the measurement should be made as a percentage of fiber content.[119]

In the absence of any other national standard, many mills have adopted the recommendations of the EPA Guideline governing federal purchases as a benchmark for defining recycled content standards in the formulation and sale of their papers to other customers. This situation has provided them

[119] "What Exactly is Recycled Paper?," *Recycled Paper News,* Vol. 1, No. 6 (February 1991), p. 6.

Table 4.1
EPA Recommended Minimum Content Standards for Recycled Paper and Paper Products

Recycled content percentages are based on total fiber weight rather than total paper weight.

Paper Product	Minimum percentage of recovered materials	Minimum percentage of postconsumer materials	Minimum percentage of wastepaper
High-grade Bleached Printing and Writing Paper			
Offset Printing Paper			50%
Mimeo and Duplicator Paper			50%
Writing Paper (e.g., stationery)			50%
Office Paper			50%
High-speed Copier Papers			50%*
Envelopes			50%
Forms Bond, Computer Paper, and Carbonless			50%*
Book Papers			50%
Bond Papers			50%
Ledger Papers			50%
Cover Stock			50%
Cotton Fiber Papers	25% (recovered cotton fiber) **and**		50%**
Newsprint		40%	
Tissue Products			
Toilet Tissue		20%	
Paper Towels		40%	
Paper Napkins		30%	
Facial Tissue		5%	
Doilies		40%	
Industrial Wipers		0%	
Unbleached Packaging			
Corrugated Boxes		35%	
Fiber Boxes		35%	
Brown Papers (e.g., bags and wrappers)		5%	
Recycled Paperboard			
Folding Cartons and other recycled paperboard		80%	
Pad Backing		90%	

* Amended from 0% to 50% wastepaper by EPA Procurement Guideline Advisory, November 1990.
** Amended from 25% recovered cotton fiber only, to include 50% wastepaper in addition, November 1990.

with what is, at least in part, a marketing strategy: Dry mill paper wastes can be saved up for inclusion in a few particular grades that can then be termed recycled. Many paper mill personnel will readily acknowledge that these wastes furnish at least some of their recycled content. Occasionally they will acknowledge that, for some grades, these wastes provide *all* the recycled content.[120]

120
Extensive conversations with mill personnel and paper distributors at many different recycling, deinking, and paper conferences throughout 1990 and 1991.

The inclusion of these wastes in much of what is now being sold as "recycled paper" has outraged many buyers and led to growing pressure for a deinked or postconsumer content standard, as well as for the development of a national program to verify manufacturers' claims. The EPA also announced in the fall of 1990 that it was considering revisions to the Guideline for printing and writing papers. In particular, the Agency invited comment on the inclusion of mill-generated wastes in their definition of

Paper is manufactured in giant rolls which are wound onto a reel at the end of the paper machine. These rolls are then transferred to a converting operation where they will be trimmed down into smaller web rolls and cut sheet sizes. Under the EPA Guideline, wastes generated during these paper mill converting operations can be used to supply the "recycled" content of new papers. (Photo courtesy of Beloit Corporation.)

wastepaper and on the possible implementation of minimum deinked or postconsumer content requirements. The Agency is expected to announce proposed revisions to the Guideline in the spring of 1992. Before any proposed changes are adopted, there will be a period for public comment.

Many purchasers would like to narrow the category of preconsumer paper wastes established under the EPA Guideline, limiting it to wastes generated after the paper has actually been sold by the mill that first manufactured it. Some states, as well as some private organizations that certify recycled papers, do not allow any paper mill manufacturing wastes to be included in their minimum standards for recycled content. Wastepaper that has been generated by independent envelope converters, printers, or other manufacturers is now sometimes called *postindustrial, postmill,* or *secondary waste,* regardless of whether it consists of printed or unprinted paper trimmings, to indicate that it has been collected from an independent manufacturing operation.

The EPA has also come under fire for a decision effective May 1990 that allows wood pulp made from sawdust to sometimes count toward the recycled content of printing and writing papers. Paper is often made from by-products of wood-harvesting operations such as wood chips, shavings, and sawdust, so it is not surprising that many people do not believe that this manufacturing practice fits the understood definition of recycling. In addition, this particular EPA decision was written to apply only to a process used by two particular companies in Maine— the Lincoln Pulp and Paper Company, which manufactures the pulp from sawdust, and Eastern Fine Paper, which subsequently makes paper from this pulp. The Agency singled out these companies with the rationale that "Maine does not appear to have viable alternative markets for sawdust."[121] The sawdust issue has raised such a furor that the EPA continues to get comments, from both consumers and industry, that criticize this decision for its inconsistency, and its irrelevance to the goal of encouraging the recycling of paper wastes.[122]

The Resource Conservation and Recovery Act is also due for reauthorization by Congress in 1992, and several bills have been introduced in both the Senate and the House of Representatives that could have a direct bearing on future definitions, standards, and purchasing requirements. Despite the broader mandates of RCRA, interest and debate have been focused on the paper issue in particular.

---

[121] *Federal Register,* Vol. 55, No. 192 (October 3, 1990), p. 40389.

[122] "EPA to Revise Paper Procurement Guideline," *Recycled Paper News,* Vol. 1, No. 8 (April 1991), p. 3.

The Recycling Symbol

The ubiquitous "chasing arrows" symbol associated with recycling has become one of the most widely used and recognized icons in our culture. Its success as a design comes from the clarity and simplicity with which it signifies the essential idea of recycling and the three necessary steps of the loop: materials collection, manufacturing into new products, and the subsequent sale and use of these recycled products. Ironically, the symbol's success is also coupled with its lack of specific meaning. It can be used by anyone, anywhere, without restriction, and indeed it has been very widely adopted. It is now used by many different industries as well as by many manufacturers within the paper industry. The mark exists in the public domain, and because it is no one's property, there are no legally binding standards governing its use. *There is no assurance that products displaying the mark have the recycled content advertised, or that they even have any recycled content at all.*

Several different organizations, including the American Paper Institute and various paper merchants, are currently publishing their own recommendations to advise people about use of this symbol. Their advice varies a great deal, and these organizations are not necessarily consistent with each other in their recommendations. Nor do any of them have the legal authority to regulate the symbol's use. So while seeing it on what we buy might make us feel good, it gives us no guarantee that we have really made the right choice.

Figure 4.2
Symbol Use as Recommended by the American Paper Institute

This organization suggests using the symbol in the two forms shown here. Their recommendations are vague, however, since no exact criteria are defined for paper products that are "recycled" or "recyclable." API also suggests using the "recycled" version of the symbol to promote the concept of recycling as well as organizations involved in recycling.

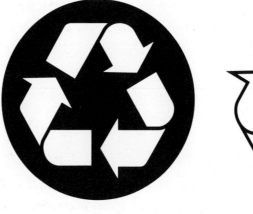
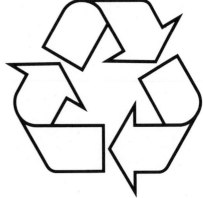

Recycled **Recyclable**

A Symbol History

The recycling symbol was designed in 1970, the year of the first Earth Day, as a public relations tool to promote the paperboard industry, which already had a record of using recycled wastepaper going back to early in the century. The Recycled Paperboard Division of the American Paper Institute asked the Container Corporation of America (CCA) to design an appropriate mark to be used on containerboard packaging. At CCA, Anthony Marcin, then manager of public relations, and William Lloyd, manager of design, organized a nationwide competition for graphic design students to develop the symbol. Several hundred entries were received, and the competition was judged at the Aspen Design Conference by several famous designers including Saul Bass, Herbert Bayer, and Eliot Noyes. The winning entry was a pen and ink drawing that was subsequently modified into a more graphic symbol by the design department at CCA under the direction of William Lloyd. Complete records of this material have since been lost, but Mr. Lloyd recalls that the winner was a student at the University of California named Gary Anderson. CCA held a large press conference in Chicago to announce the new mark, at which Mr. Anderson was recognized as the original designer.[123]

CCA subsequently applied to the U.S. Patent Office to register the symbol as a service mark, while simultaneously beginning to license use of the symbol to industry groups such as the American Paper Institute and the Corrugated Box Manufacturers Association. Their extensive promotion led quickly to its widespread use on containerboard packaging. Meanwhile, the registration of the service mark was challenged, and rather than respond to the challenge, the corporation decided to drop its application, effectively allowing the symbol to enter the public domain. Thereafter, everyone was free to use it however they chose. In recent years use of the symbol has become very widespread, and without a legal owner to define the standards under which the mark can be applied, the symbol has come to represent an idea more than a tangible reality.

State Laws and Regulations

All fifty states and the District of Columbia currently have legislation or regulations with some direct bearing on the purchase and sale of recycled paper. There is a great deal of variation from state to state in the form of these mandates. Most are legislated, but some are executive orders or other programs. The majority of the laws govern paper purchases by state agencies; as of September 1991, there were 27 state governments that

123
William Lloyd and Anthony Marcin, telephone interviews, April 1991. I have tried to locate Mr. Anderson in the hope of recognizing and reproducing his original pen-and-ink drawing, but unfortunately without success.

Below is the graphic interpretation of the symbol as refined by the design department at CCA in 1970. The bold clarity of this earlier rendition is often lost in current reproductions of the symbol, since it has entered the public domain.

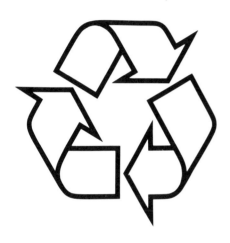

offered a price preference for the purchase of recycled paper. In addition, several states are beginning to regulate advertising claims and content standards for all paper sales within the state, especially for papers marketed as "recycled." Some states also offer financial assistance or tax credits aimed at encouraging the development and manufacture of more recycled products.[124]

In addition to the plethora of differing state regulations, a variety of regional and local governments have instituted their own legislation or programs in this arena as well. This myriad of different standards makes it increasingly complicated for the mills to do business, since they are set up to manufacture large runs of paper that can be sold to customers throughout the United States and internationally. Because some state laws or standards are stricter than others, mills face a quandary. In order to meet the tougher standards, mills must either produce their entire run of recycled paper to meet the strictest standard, or manufacture and ship special lots of paper for sale in particular states. Because the paper distribution network is primarily national, special shipments to one state can require a lot of extra tracking and paperwork.

As examples of the tougher state laws emerging on the subject, legislation from two states is discussed here. Local regulations vary widely because state or regional laws may apply and laws are frequently changed. Readers are advised to review the status of the applicable laws in their own jurisdictions.

California

The State of California promotes the purchase of recycled papers by requiring its own government agencies to offer a price preference for the purchase of recycled paper instead of virgin paper. The California price preference is 5%. In reviewing competing bids for the state purchases of paper, agencies will purchase recycled paper whenever possible, and will select the recycled product even if it is priced up to 5% over the lowest-priced virgin product. Price preferences are not unusual for state purchasing, but what is most interesting about the California law is the definition of recycled paper used for the state purchasing requirements— it is one of the strictest in the country. To qualify as recycled, printing and writing papers must contain a minimum "secondary waste" content of *50% of the total paper weight,* rather than 50% of the total fiber content. In addition, 10% of the total paper weight must come from postconsumer waste.[125]

124
"All States Now Favor Recycled Paper Products," *Recycled Paper News,* Vol. 2, No. 2 (September 1991), pp. 1–5.

125
See the section *Recycled Content Percentages* in chapter 3 for a discussion of percentages by total paper weight vs. by total fiber content.

California deliberately uses the term "secondary waste" to describe recycled content, and this differs from the term "wastepaper" that is so loosely defined under the EPA Guideline. The state's definition of secondary waste includes preconsumer and postconsumer materials, and the postconsumer definition is consistent with the EPA's. However, materials allowed under California's definition of secondary waste do not include any wastepaper generated during the paper manufacturing process itself, such as trimmings, unused paper rolls or inventories, or dry or wet mill broke. Nor does the California definition of secondary waste allow any use of wood or forest residues, wood chips, or sawdust.[126]

Effective January 1, 1991, California also adopted an environmental labeling law governing a variety of terms being used to market consumer goods within the state. Vendors and companies can only use the label or reference "recycled" for a product if it contains a minimum postconsumer content equal to 10% of total product weight. Thus printing and writing papers qualify as "recycled" under this statute if they meet the postconsumer requirement, but no additional secondary content is required. A weakness in the law is that, as written, it has been interpreted to apply only to retail sales of consumer goods, and thus it does not apply to the wholesale purchase of paper by printers, publishers, and other manufacturers. In addition, there is currently no state agency with the budget or mandate to monitor or verify manufacturers' claims, although the state attorney general is authorized to investigate and prosecute violations.

New York State
The State of New York also has purchasing requirements for its state agencies, with a 10% price preference for recycled paper. But what distinguishes New York from other states is the lead it has taken in regulating the marketing claims of manufacturers who sell their papers on the merits of their recycled content. In December 1990 a comprehensive program was instituted to regulate the sale of all products sold within the state as "recycled," "recyclable," or "reusable." Effective June 14, 1992, the use of these terms in promotion and advertising will be restricted to products that meet specific state definitions and standards for recycled content, recyclability, or reusability. Concurrently, the state licenses service marks to be used for each of those claims, and has instituted a formal application and review process under the Department of Environmental Conservation to authorize use of the appropriate service marks.

---

[126] Susan Kinsella, Californians Against Waste Foundation, telephone interview, April 1991.

Figure 4.3
New York State Service Marks
These service marks may only be used by manufacturers or companies that have been licensed by the state, and for products that meet the appropriate standards. (Reproduced by permission of NYS Department of Environmental Conservation.)

Under New York's law, paper mills and merchants will no longer be allowed to sell paper as "recycled," to any private-sector or government customer in the state, unless it meets the state standards. New York, like California, avoids use of the term "wastepaper," and it defines recycled content as being derived from "secondary material." For printing and writing papers, the minimum secondary materials content must provide *50% of the total paper weight*. No forest or wood residues, sawdust, dry or wet mill broke, or other paper wastes generated by the mill or within the same parent company count as secondary materials. Currently, for printing and writing papers, there is no requirement for postconsumer wastes to supply any of the secondary materials content, but effective January 1, 1994, there will be a requirement for a minimum postconsumer waste content equal to 10% of the total paper weight.

Participation in New York's labeling program is voluntary. If manufacturers wish to sell their paper as "recycled," however, they must apply for use of the state's licensed service mark. The application requires an executive officer of each corporation to state, in writing, that the products for which the application is made meet the minimum content requirements and that the company agrees to provide corroborative data on request in order to verify the claims. The state may revoke authorization if it determines that a product does not meet the stated requirements.[127]

Toward a National Standard

The lack of a consistent national standard requiring a specified minimum content for recycled paper *and* the lack of universally understood and accepted definitions for the many terms associated with this issue (wastepaper, preconsumer, postconsumer, mill broke, postindustrial, postcommercial, postmill, and so on) have created chaos for purchasers as well as for paper producers. In order to have any fair basis for comparison of

127
New York State Department of Environmental Conservation, *6 NYCRR Part 368: Recycling Emblems, Effective December 14, 1990* (Albany, NY: 1990).

the many papers being promoted as "recycled," paper specifiers must spend long hours asking very specific questions about the source of the wastepaper used in manufacturing, and they have to find a mill representative or paper distributor with enough specific knowledge of the mill's production process to answer these questions. This is not always an easy task.

On the other side, mill executives need to make long-term decisions about multimillion dollar equipment investments and operating costs. Even without significant changes in operating technology, it can take six months to a year of research and trials to develop a new grade of paper. With manufacturing changes such as the installation of a deinking plant, it can take three or more years to bring new products to market. Making these decisions poses difficult problems when there is no clear and consistent national standard and when purchasing regulations and directives vary widely across the country. For those mill executives who actively want to make the investments and changes necessary to use more wastepaper for their fiber furnish, a national standard would create a clear framework for decision making. But for the mill sales departments that simply want to use the recycled paper issue as a marketing opportunity, the current situation leaves a fair amount of "wiggle room."

If buyers are to be readily able to make meaningful choices between the many papers on the market, then a national standard is clearly needed. Several attempts are underway in that direction both within our government as well as within the private sector.

NASPO and ASTM

In April 1990 the National Association of State Purchasing Officials (NASPO) began a process to develop voluntary national standards for the content and definitions of recycled paper,to be used uniformly by all state governments. NASPO initiated their effort in cooperation with the American Society of Testing and Materials (ASTM).[128] They chose to work with ASTM because this organization has an established process for writing materials specifications for many different industries. ASTM procedures rely heavily on industry involvement, on the premise that manufacturers must support the standards developed if they are to be widely adopted. Within the ASTM structure and the existing Committee on Paper and Paper Products, a new Subcommittee on Recycled Paper was formed. This subcommittee was charged with developing terminology definitions and recycled content standards for different types of paper and paper products.

[128] The Council of State Governments and The National Association of State Purchasing Officials, *A Proposal to Develop Nationwide Standards for Recycled Paper* (Lexington, KY: 1990).

The ASTM committee process relies on meetings conducted over one or two days, held every few months, in different cities all over the country. In between meetings, work is done primarily by correspondence and voting on written ballots. The expense required to attend meetings sometimes limits the participation of end users on the subcommittee, and it has often resulted in meetings more heavily attended by industry representatives than by end users. But it is at these meetings that terminology is actually developed, and written ballots submitted by mail may be ruled "nonpersuasive" and thus discounted. While ASTM voting procedures are theoretically designed to balance the number of voting producers with nonproducers, the structure of the process and the difficulties posed by the limited participation of end users, have raised questions about whether this forum will be able to achieve enough real consensus to develop definitions and standards that are acceptable to all participants— the paper industry, state and federal government purchasers, other end users, and the general public.[129]

NASPO staff members have been frustrated by the slow progress of the committee work to date. The debate generated within this group, even about basic definitions, has been extensive and not easily resolved. There is no fixed completion date for the committee's work, and this poses difficulties for NASPO and other end users on the committee whose funds are limited.[130] Thus some participants are concerned that the subcommittee's final definitions and standards will be the ones most strongly promoted by a block of producers in the paper industry who have the financial resources to continue the process over a long time period. Yet there are some members of the subcommittee who are working hard to see that many different viewpoints are heard, and the outcome of this process remains uncertain.

The Recycling Advisory Council
In 1989 the Recycling Advisory Council (RAC) was formed by the National Recycling Coalition (NRC), under a grant from the U.S. Environmental Protection Agency. The RAC members were selected by the NRC Board of Directors, and several working committees were formed. The Recycled Paper Committee has begun to issue recommendations for proposed definitions and standards for recycled paper. In June 1991, this committee proposed a minimum content standard for printing and writing papers that would require 50% of the total fiber content to be recycled; concurrently 20% of the total fiber content would have to be "processed secondary fiber." RAC's proposed definition of recycled fiber was "all recovered

---

129
Attendance at ASTM Subcommittee D06.40 on Recycled Paper, organizational meeting, April 17, 1990; review of subcommittee meeting minutes and ballots in 1990 and 1991; interviews with subcommittee participants from both the producer and nonproducer categories; discussion and correspondence with Daniel Mulligan, Subcommittee Chairman, October 1991.

130
Linda Carroll, Staff Director, National Association of State Purchasing Officials, telephone interviews, May and September 1991.

paper from all sources except the virgin component of mill broke."[131] This proposed definition has been criticized because it is confusing even to those who are quite knowledgeable about mill operating procedures. In addition, by including some varying percentage of wet mill broke, it allows mills to use an even broader range of paper mill wastes than are allowed under the original EPA Guideline. In their June proposal the RAC also backed off from an earlier proposal to require either a minimum postconsumer or deinked fiber content. They decided to recommend a "processed secondary fiber" content instead. This approach was intended to recognize the benefits of using deinking grades of wastepaper without actually deinking it, as is possible for some papers. Again, concerns have been expressed by advocates in the recycling community because this definition of processed secondary fiber can be interpreted quite broadly, and it may result in the inclusion of wastes that are almost equivalent to pulp substitutes. In addition, given that many recycled papers are already made in which at least 50% of the fiber comes from deinked wastes, these critics do not believe that a requirement for only 20% "processed secondary fiber" is strict enough.[132]

In late 1991, the RAC proposals came under strong enough criticism that the Council made a decision to solicit public comments about their proposed definitions and standards, and to delay until 1992 a vote on whether to adopt them. Unfortunately, the time allotted for this review process was quite limited, requiring all interested parties to respond in a matter of only a few weeks.[133] Given the significant disagreements that exist, such a quick resolution of the issues is unlikely.

The Great Postconsumer Waste Debate

Since the advent of the EPA Guideline, paper buyers have been increasingly requesting papers with a postconsumer waste content. Simultaneously, more and more papers have been introduced with steadily growing claims of postconsumer waste percentages. The outcry for postconsumer waste has grown largely out of the frustration felt by customers who want to buy "real" recycled paper, not paper made with wastepaper discarded at the paper mill. Because the definition of "postconsumer waste" in the Guideline clearly identifies a class of wastes that genuinely require recycling through collection programs, this category of wastes has come to symbolize, in many buyers' minds, the genuine article.

---

[131] Recycling Advisory Council, "Information Bulletin" (Washington, DC: August 1991); and Victoria Ludwig, Recycling Advisory Council staff, telephone interview, September 1991.

[132] Susan Kinsella, Californians Against Waste Foundation, Memorandum to Recycling Advisory Council and National Recycling Coalition Board of Directors, June 19, 1991; discussions with other mill personnel, recycling advocates, and paper specifiers.

[133] Recycling Advisory Council, "Memorandum" (Washington, DC: October 31, 1991).

In devising an acceptable national standard, we need to resolve the issue of whether there should be a minimum requirement for postconsumer content. Given the trend toward extremely permissive guidelines for recycled content, some environmental advocates feel strongly that only a postconsumer standard will force the development of the technology and capacity needed to use the more problematic postconsumer wastes. Members of the paper industry often argue that such a postconsumer standard is unnecessary, and they do not favor requirements that mandate the use of the most problematic grades of wastepaper in the production of printing and writing papers. Occasionally, however, even a few paper executives have spoken in favor of a postconsumer content standard.

It is certainly logical for the highest grades of paper wastes (pulp substitutes and high-grade deinking wastes) to be used first for the production of printing and writing papers, since these products are more demanding to manufacture than are newsprint, tissue, or containerboard. Many of these wastes are preconsumer material, yet they are, indeed, legitimate wastes that would be landfilled or incinerated if they could not be sold to waste dealers for recycling. This is especially true of industrial waste collected from printing plants and the like. It is easy to understand the paper industry's desire to direct these wastes first toward printing and writing papers, and their corresponding reluctance to embrace a postconsumer content requirement. It is much harder for buyers to be sympathetic, however, when industry executives simultaneously lobby for national definitions and standards that allow mill-generated paper wastes to count toward minimum standards for defining a recycled product, especially considering the extent to which these materials already supply the recycled content of printing and writing papers. In addition, the fact remains that because printing and writing papers represent such a significant portion of paper consumption generally, increased use of postconsumer recycled fiber will be required if we are to develop markets for the growing quantities of paper wastes we are collecting.

Is a Postconsumer Standard Enforceable?
Probably the strongest argument against a postconsumer standard is that it may be unenforceable. Ironically, by the middle of 1991, many of the papers being sold with the highest postconsumer content claims were made by paper mills that did not have their own deinking operations. The majority of the mills that are buying deinked pulp from outside suppliers, including mills that have said they are getting pulp from Ponderosa Fibres

of America, count 100% of their purchased deinked fiber as postconsumer content.[134] Yet Ponderosa, the country's principal supplier of deinked fiber to printing and writing mills, uses on average 80% postindustrial waste and only 20% postconsumer waste.[135] Ponderosa does report that it makes an effort to include a higher percentage of postconsumer wastes in the pulp it sells to printing and writing mills, but since these mills are now purchasing over one-half of the deinked pulp that Ponderosa manufactures, they cannot all be getting pulp made from 100% postconsumer waste. The equation just does not balance.

The president of a large wastepaper and brokerage firm also spoke frankly on this matter at a paper recycling conference in 1990, stating that a mandated postconsumer content is "totally unenforceable, and an invitation for fraud, dishonesty, and abuse of the intent of the regulations."[136] He went on to describe how waste dealers will call certain grades of wastes postconsumer when they are in fact preconsumer, because in some situations it is difficult for even experienced wastepaper buyers at mills to verify any difference. A buyer purchasing shredded book paper, for example, may not be able to tell if the paper stock originally came from schools (postconsumer) or from discards at a printing plant (preconsumer).

It is difficult or virtually impossible to trace particular wastes from their source to their end use in fine paper. There is no test that distinguishes recycled fiber from virgin, much less reveals whether the source of recycled fiber was preconsumer or postconsumer. Because most mills currently making recycled paper do not do their own deinking, one would have to trace any deinked pulp they are buying several steps backward to the original pulp supplier before even inquiring about the source of the waste itself. And their supplier is probably using a whole range of wastes, pre- and postconsumer. Which wastes went into which lots of pulp shipped to which mill? If it ever becomes possible to trace the source of the wastes accurately, it certainly will require a great deal of detailed paperwork and auditing to verify the claims. Without verification, and with the high stakes of "green marketing," will all producers be honest?

Some people have proposed a deinked standard in lieu of a postconsumer standard. Since not all printed paper wastes are deinked prior to use, such a standard might refer to "deinking grade wastes" or, more simply "deinking wastes." Certainly, it is easier to distinguish between printed and unprinted wastes than between pre- and postconsumer wastes. And a mill will not

---

[134] Based on interviews of mill personnel at 50 mills, nationwide.

[135] Jerome Goodman, Executive Vice-President, Ponderosa Fibres of America, question and answers following "Wastepaper Deinking for Fine Paper Production," paper presented at *Pulp & Paper Wastepaper II Conference,* Chicago, IL (May 1991).

[136] Bruce Fleming, "A Paperstock Broker Speaks Out: Second Thoughts on Government Mandates," *Recycled Paper News,* Vol. 1, No. 2 (September 1990), pp. 4–7.

generally mix unprinted pulp substitutes with deinking wastes in an operating deinking plant, because the cost would be prohibitive. The further question that is raised here, however, is: How should the unprinted wastes that are sold for recycling by printing plants, independent envelope converters, and others be treated? Why should these unprinted paper wastes be distinguished from the printed variety? Some outspoken recycling proponents feel that this distinction is needed, but others feel that it is not. They argue that setting an enforceable standard for recycled content that is high enough, and prohibiting the use of mill wastes, makes such a distinction superfluous.

The National Regulation of Marketing Claims

The proliferation of exaggerated and oversimplified marketing claims in the environmental arena has already drawn the attention of the state attorneys general, the Congress, the Federal Trade Commission, environmental groups, members of the public, and a variety of other private-sector organizations. A task force representing the offices of attorneys general from eleven different states was formed in 1989 to address these issues. After holding hearings and reviewing testimony and comment on its initial report, this group issued its final recommendations in May 1991 in *The Green Report II*. Their report begins by stating, "Consumer clout can become a major motivating force for improving environmental policies only if the public receives accurate, specific, and complete information about the environmental effects of the goods and services they buy. Unfortunately, attempts to take advantage of consumers' increasing interest in the environment have led some companies to make environmental advertising claims that are trivial, confusing, and misleading."[137] Regarding the use of the term "recycled" to describe products, the task force recognizes an urgent need for national standards. The group also recommends clearly differentiating between preconsumer and postconsumer materials in listing recycled content, suggesting that only postconsumer materials should be called "recycled," while preconsumer industrial manufacturing wastes could be called "reprocessed [or recovered] industrial material." They further propose that by-products of the original manufacturing process (e.g., wet and dry mill broke) that are routinely fed back into production should not count as either "recycled" or "recovered" material. In addition, while they feel that sawdust is an appropriate wood by-product to use in paper manufacturing, they do not support counting its use as part of a paper's recycled content.[138]

137
California, Florida, Massachusetts, Minnesota, Missouri, New York, Tennessee, Texas, Utah, Washington, Wisconsin Attorneys General, *The Green Report II: Recommendations for Responsible Environmental Advertising*, May 1991, p. v.

138
Ibid., pp. 8–11.

Figure 4.4
Wastepaper Sources Used in the Manufacture of Recycled Paper

The debate over a national standard revolves around the types of wastepaper included in the definition of "recycled paper." Some governments and organizations use a standard stricter than that of the EPA Guideline, by requiring that minimum recycled fiber content be supplied only through the use of postmill wastes. Some recycling advocates feel that a separate postconsumer content requirement is needed in addition.

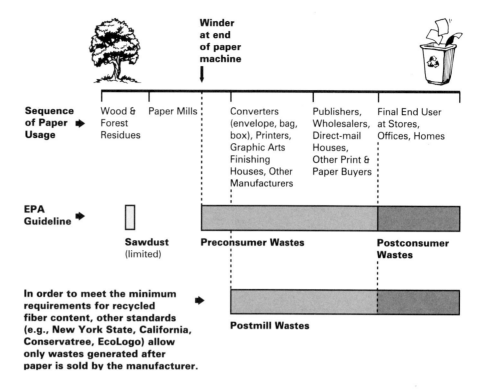

Increasingly, the Federal Trade Commission (FTC) is being asked to participate in the development of national standards regulating environmental marketing claims. The FTC has been petitioned to do so by a coalition of eleven trade industry groups headed by the National Food Processors Association. At hearings held by the FTC in the summer of 1991, the Direct Marketing Association testified to the need for guidelines governing environmental marketing claims. In *The Green Report II*, the state attorneys general urged cooperation between the states, the EPA, the FTC, and the U.S. Office of Consumer Affairs in the development of a national regulatory scheme. The FTC continues to evaluate the appropriateness of assuming a role in such matters, and the Commission is expected to announce its intentions in 1992.[139]

Private Certifications and Product Evaluations

A number of private groups have begun to develop their own programs for evaluating the environmental merits of particular classes of products or for certifying manufacturers' environmental claims. To some extent these are modeled on the successful government-sponsored environmental seal programs of other countries such as Canada, Germany, and Japan. Several difficulties are posed by such a private sector approach and the state attorneys general discuss some of the more obvious such problems in

[139] National Food Processors Association, *Petition for Industry Guides for Environmental Claims Under Section 5 of the Federal Trade Commission Act* (February 14, 1991); "Direct Marketers Lobby for Uniform Labeling Guidelines from FTC," *Paper Recycler*, Vol. 2, No. 8 (August 1991), pp. 5–6; Lee Peeler and Michael Delborrello, Federal Trade Commission, telephone interviews, April and September 1991.

The Green Report II. There is a risk that the criteria used for evaluating products will not be carefully developed, or that the exact basis on which such a seal is awarded may not be made clear to the public. Because such programs generally depend on substantial fees paid by manufacturers, not all available products are evaluated for comparison to begin with, and there is a legitimate concern about whether manufacturers who are willing to pay for certification might have undue influence on evaluative criteria. The report concludes its discussion of this matter by saying, "the Task Force sees a serious potential for deception unless certification programs are designed, promoted, and monitored very carefully. If properly implemented, certifications may offer real benefits, but opportunities for missteps abound. Manufacturers and seal grantors alike should therefore proceed with great caution."[140]

140
The Green Report II, pp. 13–17.

All paper products certified by Green Seal will meet uniform standards established for that particular type of paper; manufacturers will be able to use this symbol to identify licensed products. Green Seal has begun licensing toilet and facial tissue, and expects to start licensing printing and writing papers by the end of 1992. (Reproduced by permission of Green Seal.)

Green Seal

One of the developers of such a private seal approach is Green Seal, an independent nonprofit organization founded in 1990 by Denis Hayes, formerly the chair of Earth Day 1990. Green Seal's purpose is to develop and implement a product labeling and certification program that will recognize products that are preferable from the point of view of their overall environmental impact on the planet. The evaluative criteria will encompass the materials used to make the product as well as the methods used in the associated manufacturing process and the impact of the product after its useful life is completed. This organization has received the majority of its initial funding from private foundations, including several whose support of environmental issues is significant and well established. Additional support has come from individual donors. As Green Seal begins to evaluate particular products, however, an increasing share of their program revenues will be generated by the licensing process. The organization will charge a fee to manufacturers upon application for the seal, as well as an annual fee for continued use.

Green Seal has taken a relatively slow and careful approach in establishing the specific criteria by which each type of product will be evaluated. It has conducted an open, multistep process, with periods for comment by public and industry at several stages— before the criteria are first proposed as well as after draft criteria have been developed. In assessing paper products, Green Seal started with toilet and facial tissue. Proposed evaluative criteria were first issued in June 1991, and final criteria were completed by

November. Licensing of these products will begin early in 1992. Among other things, tissue products will be evaluated for their recycled content, bleaching process, and the level of pollution control exercised during manufacturing. Green Seal has also started to develop evaluative criteria for printing and writing papers. It expects to issue a draft of the proposed criteria in 1992, and, after several months for comment and review, to develop final criteria so that it can also begin licensing printing and writing papers this same year. Green Seal expects to use site visits to verify producers' claims, and plans to review the evaluative criteria established for each type of product at three-year intervals.[141]

Green Cross

Green Cross was also formed in 1990. It promotes itself as a nonprofit division of an established profit-making corporation, Scientific Certification Systems. However, a 1991 report by the Environmental Defense Fund questions this claim of nonprofit status on the basis that Green Cross's corporate structure does not appear to be distinct from that of its parent company and Green Cross is not registered with the IRS as a tax-exempt organization.[142] In contrast to Green Seal, Green Cross has moved very quickly to certify a wide variety of environmental claims being made in the marketing of many different products— including the recycled content of plastic, glass, steel, and paper goods as well as claims for biodegradability and energy efficiency. Within just over a year of its founding, the organization had certified over 400 different products. Green Cross does not establish a single uniform standard to be met by all products in the same category. Instead, the company will verify a specific claim for a particular product, and each manufacturer is then issued artwork for this certification, wherein the Green Cross symbol is accompanied by text explaining what is being certified in this particular case.

[141] Janet Hughes, Green Seal staff, telephone interviews, May and September 1991; Green Seal, *Proposed Criteria and Standard for Toilet & Facial Tissue* (June 17, 1991); other Green Seal documents.

[142] R. Justin Smith and Richard A. Denison, "At Cross Purposes? A Critical Examination of Green Cross' Environmental Record," unpublished report of the Environmental Defense Fund submitted to the Federal Trade Commission, the Environmental Protection Agency, and the State Attorneys General Task Force on Environmental Claims (September 30, 1991), pp. 28–31.

Green Cross does not establish a uniform standard to be met by all printing and writing papers that it certifies. Instead, labels state what portion of the recycled content is certified for that particular product.

143

Linda Brown, Jim Knutzon, and Bill Smart, Green Cross staff, telephone interviews, May and September 1991; Green Cross press kit of promotional materials, including their document *Green Cross Recycled Content Certification Standards;* additional Green Cross press releases.

144

Amazingly, in the fights over definitions being conducted in a variety of arenas nationwide, debates have even been held over how to define a "consumer." Some proposals have suggested that a consumer is "any user of an end product." Such language may be construed as effectively allowing the inclusion of manufacturing wastes not previously defined as postconsumer under the original EPA Guidelines.

Green Cross has begun to review claims of recycled content for printing and writing papers. The company's process requires paper manufacturers who apply for certification to pay an initial fee— typically at least several thousand dollars— in order to cover the cost of a site visit to audit the mill and complete certification. Additional quarterly fees are charged to maintain certification and review records submitted under their quarterly reporting requirements. The exact certification of recycled content varies with the different mills and papers evaluated, and some papers have been certified only for postconsumer content. Green Cross's definition of postconsumer materials is "a product and/or packaging material which has been discarded by an individual, commercial enterprise, or other public or private entity after having fulfilled its intended application or use."[143] This is not precisely consistent with the EPA postconsumer definition, because it does not specify wastes having reached their end uses *as consumer items.*[144]

Conservatree Paper Company

The Conservatree Paper Company, a national distributor of recycled papers based in San Francisco, was founded in 1976. This company is well known for the leadership role it has played in promoting the sale of recycled papers through the past 15 years, as well as for its lobbying efforts for a national standard and for stricter EPA definitions. Conservatree does not manufacture its own paper; it is a distributor, and it contracts with other mills to have papers made, many of which it then sells under its own brand name. The company ranks the papers it sells according to its own standards for recycled content. This rating system has four categories: C_3, C_2, C_1, and the highest rating, C_{1+}. Conservatree asks a representative of each mill to certify, in writing, the details of each paper's recycled content, and the request for information is quite specific. The company does not, however, make site visits to the mills in order to verify claims.

In 1991 Conservatree proposed an "Advance Disposal Fee" or an increase in the federal taxes to be paid by paper manufacturers in proportion to their sales of virgin papers. Coupled with this would be a "Waste Reduction Credit" or tax credit for manufacturers granted in proportion to their sale of recycled papers. The company has proposed that a rating system comparable to its own be adopted by the federal government in order to determine which papers would be eligible for the proposed waste reduction credit. Papers with higher rankings would receive more of a tax credit than papers with lower ranking.[145] Should such a proposal be adopted, the ranking system would need to be expanded to encompass all printing and writing papers sold rather than those sold only by Conservatree. And the difficult question still remains as to what independent body would have the authority and capability to establish the rankings and properly verify the manufacturers' claims. To be credible, this would be a tremendously difficult, costly, and time-consuming task.

[145] Conservatree Paper Company, *ESP News,* Vol. 4, No. 1 (March 1991); Conservatree Paper Company, *Get Real! A Consumer Guide to Real Recycled Paper,* 1990.

Table 4.5
Conservatree Paper Rankings

C_{1+}	60% postmill material by total paper weight, including 15% postconsumer material by total paper weight. (Planned increase of postconsumer content minimum to 25% of total paper weight, in 1992.)
C_1	50% postmill material by total paper weight, including 10% postconsumer material by total paper weight. (Planned increase of postconsumer content minimum to 15% of total paper weight, in 1992.)
C_2	40% deinked material by total paper weight.
C_3	50% postmill material by total paper weight.

Conservatree definitions for recycled paper do not include sawdust, forest residues, mill broke, or any paper wastes generated within a paper manufacturing operation. The three terms used in their ranking system are defined as follows:

Postmill material: Paper wastes generated during production that cannot be returned to the same production process or used by another company to make a product similar to the original product. This includes wastes generated during the intermediate steps in producing an end product by succeeding companies.

Deinked material: Printed or coated paper, the fiber of which must undergo a process during which most of the ink, filler, and other extraneous material is removed.

Postconsumer material: Only those products generated by a consumer that have served their intended end uses and have been separated or diverted from solid waste for the purpose of collection, recycling, and disposition. Wastes generated during production of an end product are excluded.

146

The words "EcoLogo" and "Environmental Choice," and the symbol below (M), are official marks of Environment Canada. Products that meet the criteria established for the program can be licensed to carry the EcoLogo symbol. (Reproduced by permission.)

The Canadian Environmental Choice Program

The Canadian government has instituted a labeling program to recognize products that meet established criteria for environmental superiority—meaning that the manufacture of these products has fewer negative environmental effects than that of some competing products. Manufacturers submit their products voluntarily, and the ones that meet the established standards can be licensed to display the EcoLogo symbol.[146] This gives customers a basis on which to make purchasing choices. The standards are developed by the Environmental Choice Board, a body of sixteen people appointed by the Minister of Environment. An independent certifying organization under contract to the federal government, the Canadian Standards Association, then reviews applications to determine which products properly meet the criteria in order to be licensed. The fees charged to applicants are exactly coupled to the time and expense incurred to review the application. An annual fee is then charged to review and maintain the license.

For recycled printing and writing papers the guidelines are straightforward and clear: *A minimum of 50% of the total paper weight must be recycled fiber, of which an amount equal to 10% of total paper weight must come from postconsumer sources.* The definitions of terms are clearly stated. "Recycled fiber" is defined to include postcommercial waste from industrial printing, converting, and manufacturing operations. "Postconsumer fiber" is derived from retail stores, office buildings, homes, or municipal collection systems, after having served its end use as a consumer item. The guideline specifically states that "recycled paper" excludes wet or dry mill broke, precommercial waste created during the converting of paper by a paper manufacturer into sheets, rolls, or other paper products (regardless of location), and all forest residues such as wood chips and sawdust. Furthermore, all manufacturers licensed to use the EcoLogo must also be in compliance with all Canadian Environmental Protection Act laws governing air and water pollution discharges.

The EcoLogo application process requires the chief executive officer of the corporation to attest in writing that the product for which it has applied for licensing meets the guidelines. The Canadian Standards Association makes a site visit to inspect procedures, record keeping, and environmental compliance. CSA is authorized to audit the company's production records and facilities and has the right to make unannounced visits to verify that the standards are being followed. Use of the EcoLogo is only licensed for the specific products that meet all the prescribed requirements, not for all the products the manufacturer makes. Wherever the symbol is displayed, prescribed language must be included stating exactly what the symbol signifies. For printing and writing papers it is always accompanied by the words "over 50% recycled paper including 10% postconsumer fiber."[147]

Though this program was developed for merchants and purchasers in Canada, the Canadian Standards Association will license mills in the United States as well. As of December 1991, six paper mills in Canada and six in the United States had had one or more of their papers certified.[148] The Canadian EcoLogo program is comparable, though not identical, to the New York State labeling regulations. In both cases any paper manufacturer may apply for rights to market papers under the registered trademark; the criteria for meeting the guidelines are well defined; no forest residues or paper mill manufacturing wastes are included in the definitions of recycled content; and consistent standards are applied to all printing and writing papers within the program, so that similar products displaying the symbol meet the same standard. The principal difference between the programs is their geographic scope. Merchants who sell recycled paper in Canada have only one clear mandate to meet for papers sold throughout the country. Interest in the EcoLogo program is growing in the United States; this program may thus begin to provide a standard for buyers throughout the North American continent.

---

[147] Canadian Department of the Environment, *Canadian Environmental Protection Act*, Section 8(1)(b).

[148] Linda Lyle, Licensing Coordinator, Canadian Standards Association, telephone interview, September 1991; Canadian Standards Association, *Information Sheet on Environmental Choice EcoLogo Licensing* and *List of Licensed Manufacturers and Products* (December 2, 1991); news bulletins published by Environment Canada.

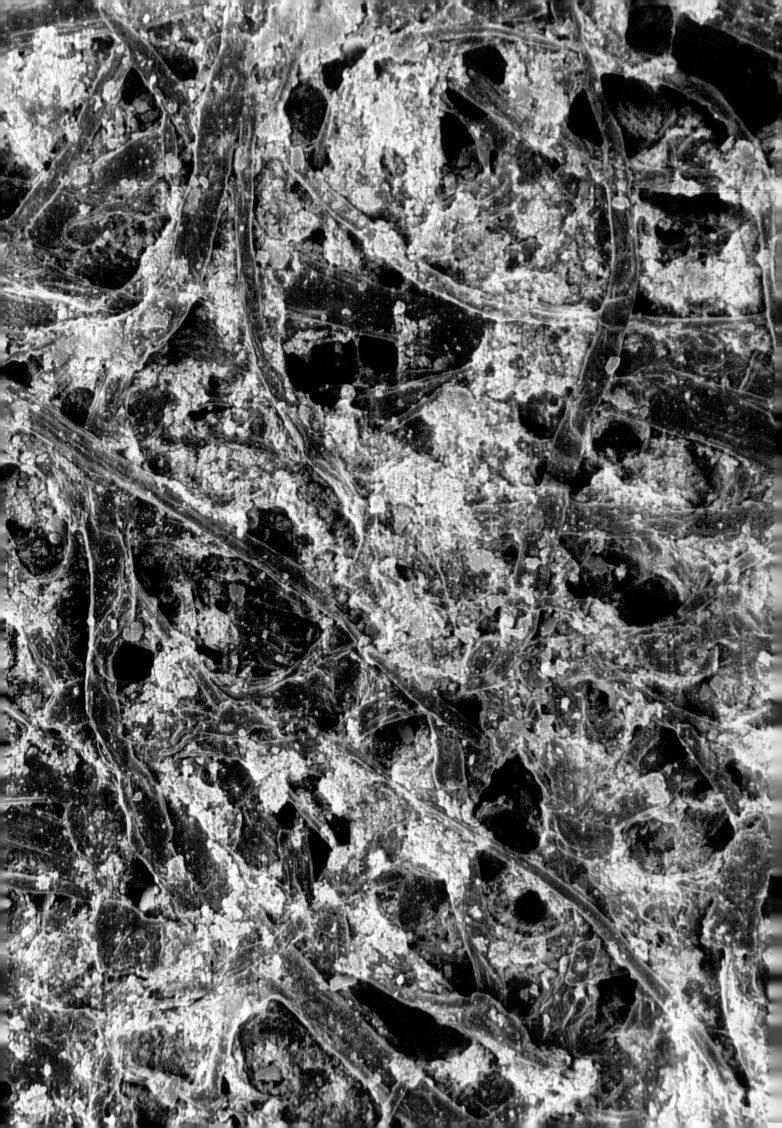

5 The Characteristics of Recycled Paper

An enlargement at 450x of recycled paper with a fiber content consisting of 50% deinked fibers and a total of 70% recycled fibers. The paper's filler content includes calcium carbonate. The photograph was taken with an electron microscope, and the image primarily shows the paper's surface.

It is very difficult to make accurate generalizations about the physical properties of recycled papers or about differences between recycled and virgin papers. For both products, many variables contribute to the ultimate quality of the paper. In virgin paper manufacturing, the particular types of wood species or other plant materials used to provide the cellulose fiber are crucial to determining the characteristics of the paper produced. The properties of recycled papers will also be greatly influenced by the fiber source— the type of wastepaper consumed, how heavily it was printed on, and how many contaminants were present, as well as by the original plant material and pulping method used to extract the cellulose fiber when it was first made into virgin paper. In both cases, the sophistication of the mill's manufacturing process and its production standards will also have great bearing on the quality of the product. Recycled papers have been made in the United States throughout this century by a few mills, and they have been sold in a competitive marketplace dominated by virgin papers. Until recently they were not often labeled as "recycled," and buyers were typically unaware of any difference in the origin of the fiber. The focus of manufacturing developments in the last century has been primarily on the use of virgin fiber. Certainly wastepaper does provide a less uniform and predictable source of fiber, and this poses clear challenges to manufacturers. Interest in the subject is unprecedented, however, and dramatic manufacturing developments are starting to take place. We are on the threshold of an era in which recycling will become the norm.

The Physical Properties of Recycled Fiber

When paper is made, the pulp is refined immediately before it is sent to the paper machine. This refining process is crucial to papermaking— it cuts the fibers to the desired length and roughens the edges of the formerly smooth fibers, raising little fibrils from the fiber wall. This increases the surface area available between the fibers, which allows the necessary chemical bonds to form. Without this fibrillation process, paper cannot be made.

The way in which papermakers refine fiber depends on several factors, including the starting fiber characteristics or original plant material from which the pulp was derived, and the type of paper product being made. For example, hardwood fiber is refined differently than softwood fiber; fiber destined for printing and writing papers will be treated quite differently than fiber destined for paper bags. Papermakers who work with recycled fiber must develop a refining treatment appropriate to the properties of this material, and recycled fiber itself can be quite variable. Its characteristics depend on the original plant fiber source, the product from which the recycled fiber is taken, and the intervening treatments to which it has been subjected.[149]

There is tremendous variability among papers generally, but perhaps especially among the recycled papers now on the market. This may be due to the rapid shift toward the manufacture of recycled papers and the fact that many mills that had been producing only virgin paper have quickly introduced new recycled grades. Inevitably there is a learning curve associated with making a product from a fiber source with which mill personnel may be less experienced. There is nothing inherently inferior about recycled cellulose fiber that has been properly processed and cleaned for reuse. Because it has already been processed once, however, it does have some characteristics that make it different from fiber derived directly from virgin pulp— compared to unrefined virgin fiber, it tends to be shorter, stiffer, and already frayed at the edges. For the papermaker, this has both advantages and disadvantages.

Advantages
Recycled fibers may yield paper with slightly more bulk (thickness) and opacity at a given basis weight than paper made with virgin fiber. Printers have also reported less of a tendency for cracking to occur in recycled paper when it is folded against the grain, sometimes less need to score the fold on heavier basis weight papers compared to virgin papers of equivalent

---

[149]
Information for this section on physical properties was compiled from a number of lengthy interviews with paper industry representatives and consultants including: Jobe Morrison, President of the Miami Paper Mill, July 1991; Charles P. Klass, Klass Associates, September 1991; production managers and personnel at paper companies including Crane Paper, Patriot Paper, Strathmore Paper, and others.

weight, and less folding problems overall. Some printers also feel that recycled paper has a little less tendency to curl and better properties for lying flat, though there is no general agreement on this particular matter. The theory behind these observations of bulk, opacity, and folding is that overall the fibers in recycled paper may be shorter than they are in comparable virgin products. Some of the shortest fibers may better fill in the spaces between the general grain direction of the longer fibers, and this would lead to both greater opacity and improved folding characteristics.[150]

For book and magazine publishers, the greater bulk and opacity of recycled papers can be an advantage. A lighter-weight paper can then be used without show-through, resulting in cost savings for the purchase of the paper as well as postage. The greater bulk can also increase the thickness of a book without increasing the weight. This is useful when publishers wish to create the aura of an impressive volume or simply want to "bulk up" a small book to make it easier to bind. The tendency of recycled papers to lay flat is also beneficial, and this attribute was used historically, long before the 1980s, as a positive sales point by representatives of recycled mills, when their product was competing in a marketplace dominated by virgin papers.[151]

Challenges

One of the biggest difficulties in working with recycled fiber is to achieve superior sheet formation. (Sheet formation may be observed by holding the paper up to light and looking at the evenness of the fiber distribution: The more perfect the sheet formation, the more evenly will light be transmitted through it.) Superior sheets are easier to print on because the ink will lay down more evenly across the entire surface of the paper. If the formation is uneven, the hills and valleys created will take the ink somewhat differently, and this can lead to printing problems.

For all papers, the quality of sheet formation and the resulting smoothness of the paper are controlled by many factors, including the degree to which the fiber has been refined, the addition of fillers that take up the spaces between the fiber, the use of coatings or surface sizings, the degree of calendering, and the type of pulp material used.[152] Sheet formation may be more difficult to control when recycled papers are manufactured for which recycled pulp is refined together with virgin pulp. This can result in the already developed recycled fiber being over-refined and the virgin

---

150
Roundtable discussion at Daniels Printing, April 1991; conversations with representatives of several other printing firms; Anderson Fraser Publishing, *Recycled Paper for Printers: Manual* (London: Anderson Fraser, 1990), p. 10; Gurudas G. Khambadkone, "Virgin vs. Deinked Pulp: A Comparison," unpublished report of the Miami Paper Mill, West Carrollton, OH).

151
Interview with Bill Walsh, Manager of Manufacturing at Patriot Paper, March 1991.

152
Anderson Fraser Publishing, *Recycled Paper for Printers: Manual,* p. 8; Charles Klass, telephone interview, September 1991.

The formation of paper is best observed by holding it up to light. Paper with an even formation will transmit light in a uniform manner. If the formation is uneven, the thicker areas will appear darker, giving it a lumpy or mottled look. Good formation is important to paper qualities such as strength and ease of printability.

Top: *Good formation.*

Bottom: *Poor formation.*

(Photos © Stanley Rowin.)

pulp not being refined enough, leading to lumpy formation. Problems also result if manufacturers do not have their papermaking chemistry in proper balance, and the use of recycled fiber with postconsumer content may require a different paper chemistry than that used with pure virgin fiber.[153] Manufacturers of recycled paper can control these factors to achieve good formation. Some also report making other modifications to their paper machines— in particular, slowing the speed of the machine or lengthening the formation table— to make their operations more compatible with the use of recycled fiber.[154]

The Difficulties of Generalizations
Papermaking is indeed an art as well as a science— an art that those of us who simply buy paper and use it, but do not make it, have a tendency to take for granted. There are so many variables that influence the final quality of the product that readers should be skeptical of oversimplified generalizations about recycled paper or any other paper. The decreased uniformity of recycled fiber recovered from a diverse waste stream, in comparison to virgin fiber obtained from known species of trees or plants, may pose challenges, making recycled paper more difficult to produce, but the successful performance of many recycled papers indicates that those challenges are being met. The best recycled papers compete quite readily with virgin papers, and many of these papers have been used for years by printers and purchasers who are happily unaware of any difference in fiber or of any problem due to manufacturing method. Indeed, my own printing experience with recycled papers, like that of many other designers across the country, has primarily yielded very satisfactory and pleasing results.[155] Since "folk" generalizations have nevertheless been made about problems associated with recycled paper, it may be helpful to review here a range of printing problems that might be encountered with any paper, coated or uncoated, virgin or recycled.

Working with Your Printer

Printing problems are caused by a wide variety of reasons that may or may not be a function of the paper at all. A good printer will work hard to anticipate and solve such problems. If a problem cannot be solved on press, and is indeed being caused by inferior or defective paper, the printer should be willing to stop the press run and make a claim to the paper merchant or manufacturer from whom the paper was purchased.

153
In all papermaking, virgin or recycled, a careful chemical balance of all materials (fiber, filler, additives, and water) must be maintained in order for proper chemical bonds to form between the fibers. Contaminants remaining in processed secondary fiber may adversely affect bonding characteristics of the fiber, requiring adjustments to the papermaking chemistry. Mills with the best cleaning and deinking operations will have an easier time working with recycled fiber.

154
Interviews with paper production managers at several recycling mills.

155
Based on interviews with principals of many design firms throughout the United States, as well as with designers in large corporations that have made major commitments to using large quantities of recycled paper.

156
Several years ago I was being interviewed for design work by a publisher of limited-edition fine-press books. I thought I recognized the particular paper several of their books had been printed on, and when I read the colophon my guess as to the identity of the paper was confirmed. "Did you know this was recycled paper made with deinked wastepaper?," I asked. The publisher looked quite surprised and gave a slightly shocked reply, "No!" They had used this particular paper for years without any problems.

157
Roundtable discussion at Daniels Printing, April 1991; interviews with several other printers.

158
Anderson Fraser Publishing, *Recycled Paper for Printers: Manual,* pp. 12–13.

159
While printers have reported this problem to be associated with recycled papers, when further questioned most state that the occurrence is not consistent. They indicate that many print runs on recycled paper go perfectly smoothly and that, furthermore, linting and picking problems also occur with virgin papers. They also report that they may experience quite different results with the same grade and type of paper— one lot may print well, while another exhibits difficulties. (Roundtable discussion at Daniels Printing, April 1991; interviews with representatives of several other printing firms.)

Not surprisingly, different printers have taken very different attitudes toward the use of recycled paper. Some press operators, when faced with difficulties on press, have a tendency to look for causes beyond their control. If the paper is labeled as "recycled," they may have a predisposition to put the blame here before searching for other possible causes requiring adjustments of press conditions or ink formulations. This can, unfortunately, lead to recycled paper becoming an excuse for a poor printing job. Yet other printers have made an enthusiastic commitment to the use of recycled papers for at least two reasons. First, they recognize that these papers have in fact existed and performed well for many years; they know that they may have printed on recycled papers that were not labeled as such, and they have been unaware of any difference.[156] They are also confident enough of their own knowledge and skills that they do not fear having to make a paper claim if it is determined that the paper, indeed, is at fault. Second, they perceive that the specification of recycled papers is increasing. From a business perspective, they want to position themselves as having capabilities and expertise in this area, in order to increase their share of the printing business.[157]

The success of any job requires a good working relationship between the client, designer, and printer. Cooperation between the designer and printer should begin in the early design phase of the project, in order to plan how the job will best be printed. An early review of design comps and job specifications will help both the designer and the printer. A good printer's representative will provide a wealth of information about problems that might be anticipated. This discussion of printing specifications should include consideration of the paper specified, since paper is a big variable in any printing job. If the product is well designed, from the point of view of choosing appropriate papers and printing processes to achieve the desired results, the likelihood of encountering problems will be diminished.

Linting and Picking Problems
"Linting" is the term used to describe the pulling of loosely bonded fibers from uncoated papers; "picking" refers to the pulling of coating or small pieces from the surface of any paper. Linting and picking can lead to hickeys, or to white specks showing in solid printed areas. Recycled papers have been reported to be prone to linting and picking.[158] The tendency of paper to lint and pick, however, is due to many factors, including the characteristics of the fiber bonds, surface sizing, and the presence of foreign matter. These problems can in fact occur with recycled or virgin papers.[159]

Linting and picking problems can result in hickeys (left photo), when loose paper fibers build up on the press plate or blanket, preventing an even transfer of ink. They also result in loss of solid ink coverage where paper fibers have pulled away from the printed area (right photo). (4U symbol designed for gift wrap, © Claudia Thompson.)

Linting and picking are more commonly associated with uncoated than coated papers. Many printers are more experienced at printing on coated papers, and standard press settings and press conditions are often set for coated rather than uncoated papers. If the necessary press adjustments appropriate to an uncoated stock are not made, problems are more likely to occur. There are many tack forces exerted on paper as it passes through the printing press, including press speed, ink tack, and even blanket type. All of these can contribute to the pulling of fiber or coating from paper. If problems occur, these tack forces can be reduced in several ways— by slowing the press speed, by reducing the tack of the ink, and even by changing the blanket type of the press to one that has a softer surface and is fully compressible.**160**

160
Anderson Fraser Publishing, *Recycled Paper for Printers: Manual,* pp. 12–16.

All paper is made to a given set of specifications that describe how it should perform. Paper mills routinely run a pick test on paper at periodic intervals during the manufacturing process. This test is not conducted differently for recycled and virgin papers; it simply measures the tendency of the fibers or paper coating to pull away from the body of the paper. If significant linting and picking problems occur on press and cannot be solved through reasonable pressroom adjustments, the paper may indeed be at fault. The particular lot of paper may not be up to the specifications. In this case, the printer should stop the job and make a claim to the paper company. Evidence of the problem should be gathered in the form of a "tape pull." This consists of simply using clear tape to lift particles of dust, lint, and fiber off the press blanket and applying this tape to a clear piece of acetate. This gives the paper supplier a visual record of the problem as it occurred.

Solid Ink Coverage

For all papers, virgin and recycled, large areas of solid ink coverage are always more difficult to print perfectly than are areas of type and line art. Usually, coated papers will give better ink hold-out for solids than will uncoated papers, though this does not mean that solids cannot be successfully printed on uncoated papers. Printing large solids can aggravate the tendency of papers to lint and pick, since the tackiness of ink printing over this large surface area may tend to pull any loose fibers off the surface of the paper. These loose fibers may also build up on the blanket, subsequently causing the formation of hickeys during the print run. In some circumstances, for virgin or recycled papers, it may be desirable to print solid areas with two passes of color over the same area, in order to cover up imperfections that might show with only one printing of the solid. It may be especially desirable to do so when printing on uncoated papers (since they are more prone to linting and picking) or when printing on papers with an uneven sheet formation.

Uncoated papers, especially if they are alkaline-sized, are also typically more absorbent than are coated or acid-sized uncoated papers. This property, along with any unevenness in sheet formation, may result in the press operator having to put a bit more pressure on the paper while it passes through the press, in order to get a uniform ink coverage in solid areas. It may also require ink coverage overall to be run a bit heavier than normal. These steps, in turn, can increase the dot gain, especially in the middle ranges of halftones.

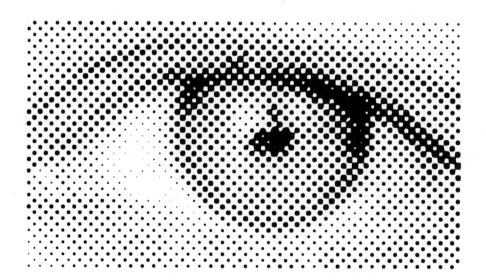

Continuous-tone photographic images are reproduced by screening or scanning the image into halftone dots, as shown in this enlargement of a screened image. During printing, some dot gain (enlargement of dot size) will take place in the transfer of ink from printing plate to paper. The amount of dot gain will be directly affected by the type of paper printed on. This can be compensated for by proper screening of the original image.

161
Anderson Fraser Publishing in England has developed a detailed guide for printers working with recycled paper that addresses the issue of dot gain in the most comprehensive manner I have seen. Part of a more complete set, this particular resource book is titled *Recycled Paper for Printers: Manual.*

162
Higham, *A Handbook of Papermaking*, pp. 152–154.

163
If the paper has a textured surface created by the use of a dandy roll, such as a laid, linen, or other textured finish, the texture will be most pronounced on the felt side of the sheet where the dandy roll impression was made. Otherwise, on smooth finish papers, the wire side will usually be very slightly rougher than the felt side, though the difference is so slight that most observers, including many printing plant personnel, may not be able to distinguish the two sides.

164
Paul Merna, Lindenmeyer Munroe (paper merchants), telephone interview, May 1991; discussions with printers and production personnel at paper mills.

Dot Gain

On almost every job, some dot gain takes place between the making of the negative and the final transfer of ink to paper. If the paper to be printed on is uncoated, and especially if it is fairly absorbent, it is necessary to adjust for dot gain. To compensate, the scanner operator who is screening and separating the photographic images should evaluate the printing process to be used and the paper specified before scanning the images, in order to gauge the expected dot gain. By then screening the images with reduced dot size, especially in the midtones, which tend to be the most affected, the resulting dot gain can be compensated for.[161]

Uncoated Papers: The Felt and Wire Sides

The two sides of uncoated papers can be quite different, due to the way paper is formed on the paper machine.[162] As the pulp travels down the moving wire screen, water is removed by suction under the wire. Once the pulp forms into a sheet of paper, it is carried from the wire between felts that pass through wringer presses; this further flattens and smooths the fibers into the paper and removes more moisture from the sheet. Subsequently, the paper passes through the long section of drying drums. The side of the paper that was originally in contact with the wire screen of the paper machine is known as the *wire side*. The other side is termed the *felt side*, even though both sides of the paper usually come in contact with felts.[163]

Some papermakers have hypothesized that the fibers on the wire side of the sheet may be more tightly bonded into the paper, with the finer fibers and filler floating more to the top or felt side of the sheet during formation. This would make the wire side less prone to linting or picking. Not all paper industry representatives agree on this point, however; others have stated that with changes in paper machine technology, and depending on the application of surface sizings or film coatings, there is less difference between the felt and wire sides of paper than there once was. Regardless, most agree that the two sides of uncoated papers may print differently. If problems are encountered in printing and the job only requires printing on one side of the sheet, it is certainly recommended that the printer turn the paper over and try the other side. For two-sided jobs, paper merchants also recommend as general procedure that, during makeready, the printer try printing on both sides of the sheet, to isolate any possible problems that might be encountered later during the print run when the paper is turned over.[164]

165
Anderson Fraser Publishing, *Recycled Paper for Printers: Manual,* pp. 10–11.

166
John Braceland, owner of Braceland Brothers Printing, interview, April 1991; roundtable at Daniels Printing, April 1991; interviews with additional printers.

Misregistration

Some printers have reported a tendency for recycled paper to stretch in both directions on press, creating a potential for misregistration.[165] All paper has a grain, and in sheet-fed printing presses paper tends to stretch more across the grain than otherwise. Proper planning by the printer will almost always result in misregistration problems being a nonissue. Practically speaking, misregistration is most likely to occur on jobs using exceptionally large sheet sizes, over 25 x 38", and it is almost always possible to run sheet-fed jobs on paper of this size or smaller. On larger sheet sizes, misregistration of multicolored jobs is more likely for both recycled and virgin papers. Critical registration areas are best positioned toward the middle and the leading edge of the press sheet, if possible, and a good printer will generally do this for all jobs regardless of the type of paper.

Web Breaks

The generalization has also been made that recycled papers made for use on large web presses are more prone to breaking under the high tension and speeds of this printing process. However, web breaks are not uncommon with any paper, whether virgin or recycled, and there is a lack of consistency in the reporting of this problem. Sometimes it is the tension and the speed of the presses that require adjustment, and not the paper that is at fault. Some printers who have experienced repeated web breaks when using recycled paper have found that their problem was solved by switching to recycled paper made by a different manufacturer.[166] Paper is made to a variety of strength specifications, which include tests for bursting strength, tensile strength, and tear. Again, the test criteria are not different for virgin and recycled papers, and any paper should perform within the specifications it is made to.

Paper Claims

Given the pattern of reporting, it seems reasonable to conclude that problems of paper strength and linting or picking are not inherent to recycled papers but are indicative of the quality standards to which the paper has been made. A responsible mill will stand behind its product and honor a paper claim if that product does not meet specifications, regardless of whether the paper is virgin or recycled.

167
The Paper Complaint Handbook System, published by Printing Complaints Incorporated, provides detailed information about typical paper problems and how to approach and handle paper claims.

168
Grade Finders Incorporated, *Competitive Grade Finder: 1990/91* (Exton, PA: Grade Finders, Inc.), p. i.

169
Most test methods use magnesium oxide as the standard reflective material against which paper is compared. Measuring equipment is calibrated to define this material as 100% reflective. Subsequent tests then compare the amount of light reflected by the paper to that reflected by the standard. Paper with a brightness value of 80 reflects 80% of the light reflected by the magnesium oxide standard.

Printers should have established procedures and forms for making claims, but they are sometimes reluctant to do this because they do not always get reimbursed for the press time already expended during the makeready or printing period before the problem occurred. Printers do have to make paper claims periodically for a wide variety of paper-related problems. In order to make a claim, the printer is expected to notify the paper merchant or mill immediately while the job is on press, to give the supplier an opportunity to help solve the problem without replacing the paper. Any claims subsequently submitted for jobs which have been pulled off press, and for which replacement of the paper is requested, must include detailed information about the paper, the lot number, the exact printing conditions, and evidence of the problem. While the merchant or supplier has the responsibility to investigate and respond to the claim promptly, the difficult side of this procedure is that it does necessitate holding up the job until the claim is resolved. A printer who has a good working relationship with its merchant should, however, be able to resolve claims fairly quickly.[167]

Brightness Standards

The classification of papers into the categories commonly used in the paper industry— "Premium," "No. 1," "No. 2," and so forth— is done "primarily through brightness and opacity [using data] furnished by the paper manufacturers."[168] These classifications have contributed to the increased sale of papers with the highest brightness, especially in the United States, and even to the mistaken belief that such papers are superior to others, even though this grading system was never intended to compare the complex attributes of overall quality and performance.

Paper brightness is a measure of the reflectivity of paper to light, with higher brightness numbers indicating greater reflectivity than lower numbers. Brightness values can hypothetically range from 0 to 100, with typical numbers for white offset printing papers falling between 70 and 90. The industry uses several established test methods for measuring brightness, and each test will yield slightly different numbers for the same paper due to minor variations in methodology for each testing procedure. Thus, in comparing the brightness numbers of different papers, one should ascertain whether the same testing procedure was used.[169] Brightness is not a measure of color, and papers with the highest brightness numbers are not always perceived by viewers to be whiter than others with lower numbers.

Because brightness tests are conducted using a single wavelength of light in the blue region of the spectrum, papers with the highest brightness values are not even necessarily more reflective of light in daylight or the typical viewing conditions in which the paper may ultimately function.[170] In ranges below 75–80 brightness, however, most people will perceive papers with higher brightness values to be whiter or lighter than those with lower values.

In papermaking, two principal factors influence brightness. First, it is a function of properties of the pulp used. Bleaching of pulp is crucial to increasing final paper brightness, but other pulp properties such as groundwood or lignin content are important as well. Second, brightness is enhanced by the addition of fillers and additives to the paper. All printing and writing papers include fillers. Clay, which is often used as a filler, adds some opacity and brightness. In alkaline papermaking, calcium carbonate ($CaCO_3$) is used as the primary filler. This is an even more effective brightening agent than clay. The brightest white papers, with values of 85–90 or higher, are manufactured with the addition of titanium dioxide (TiO_2) or fluorescent dyes. Though fluorescent dyes do not add opacity to the paper, they are highly reflective and are therefore effective brightening agents.[171]

Setting aside all the complicated variables that control the final brightness of paper, any difference between virgin and recycled products would arise from limitations in the brightness of the pulp obtainable from wastepaper rather than from virgin wood or other vegetable matter. The brightness of recycled pulp is determined by the type and age of wastepaper used, the deinking chemistry and systems employed, the water clarification system at the mill, and the bleaching method.[172] Overall, there may be a slight reduction in the brightness attainable from the processing of wastepaper rather than from the bleaching of virgin wood pulps. Wastepaper fiber may retain small amounts of dyes or ink residue, even when deinked in the best of systems.[173] Nevertheless, there are several white offset papers in which over 50% of the fiber comes from deinked wastepaper that have final brightnesses in the 80–85 range. This is a higher brightness than that of many printing, office, and business papers currently in use, and is suitable for virtually any printing job. In short, the use of recycled fiber does not preclude the production of clean, attractive, bright paper with good properties for reproducing halftones and other art.

---

[170] Casey, ed., *Pulp and Paper Chemistry and Chemical Technology,* Vol. 2 (1st ed.), pp. 888–892.

[171] The presence of fluorescent dyes can easily be detected by observing the paper under a black light.

[172] Michael Dodson and Lowell Dean, "Proper Deinking Chemistry, Bleaching Technique Crucial to Pulp Brightness," *Pulp & Paper,* Vol. 64, No. 9 (September 1990), pp. 190–191.

[173] Casey, ed., *Pulp and Paper Chemistry and Chemical Technology,* Vol. 1 (3rd ed.), pp. 578–594.

The desire for products with exceptionally high brightness levels is being reevaluated for several environmentally related reasons. First, it is reasonable to question the addition of titanium dioxide and fluorescent dyes to our paper in order to achieve a super bright white. Neither of these compounds is completely benign, and both represent potential contaminants to our water supplies if released during the papermaking process. Second, chlorine bleaching methods have traditionally yielded the brightest pulps, but the use of chlorine compounds is associated with the production of a large number of toxic chlorinated organic compounds, especially in the multistage bleaching sequences typical of chemical pulping processes using virgin wood. A complete discussion of the bleaching process itself, and the controversy over chlorine methods, would require an entire book. There seems to be a growing consensus, however, that environmentally preferable methods of bleaching should, and will, be phased into the industry. Some of these methods, using hydrogen peroxide, oxygen, ozone, or related compounds, may yield pulps that are slightly less bright, though techniques using these methods are being studied intensively and are progressing rapidly.[174] In the final analysis, and after decades of conditioning, it may behoove us to admit that the highest attainable brightness is not, in fact, what really yields the best paper.

Permanence of Recycled Paper

Papers made in Europe and America from cotton and linen rags have, in general, held up amazingly well. Many books and records made three or four centuries ago are in very good condition and can be handled and used without fear that the pages will crumble. Yet this is not true of many more recent documents, especially those printed since the advent of wood-pulping methods and the increased used of alum in papermaking. Soon after the development of groundwood pulping methods in the latter half of the nineteenth century, it was observed that paper made by this method yellowed, became brittle, and crumbled relatively quickly. Even with the subsequent development of chemical wood pulps, permanence problems persisted, until it was ascertained that the acidity of paper, created by the use of alum-rosin sizing, made it much more prone to deterioration than would be the case if it were neutral or alkaline to begin with. For this reason, since the 1950s there has been a growing conversion of paper mills from acidic to either neutral or alkaline papermaking processes.[175]

---

[174] Kimmo Järvinen, Process Engineer, Peroxygen and Pulp Chemicals, interview at *New England TAPPI/Connecticut Valley Pima Technical Seminar* (March 1991).

[175] Background information for this section on permanence has been compiled from a number of different sources: Mead Paper, Graphic Arts Knowledge Series, "Paper Permanence" (1983); Ellen McCrady, editor of the *Alkaline Paper Advocate*, telephone interview, May 1991; Chandru Shahani, Library of Congress Preservation Research Office, telephone interview, May 1991; interviews with other paper industry executives.

The factors that influence the permanence of paper are complex. It is widely agreed, however, that the two most important characteristics in determining permanence are the purity of the cellulose fiber in the pulp and the final pH (alkalinity or acidity) of the finished paper. In wood pulps it is believed that the presence of lignin is probably the most significant factor that adversely influences permanence, and all current standards for paper permanence specify that the fiber contain no groundwood content. Wood pulps that have been chemically processed to remove the lignin are termed in the industry "wood-free," and depending on the extent and conditions of the cooking process during pulping, more or less of the lignin will be removed. Accepted industry standards actually define "free-sheet" papers as those that contain no more than 10% groundwood pulp.[176] Since cotton contains virtually no lignin to begin with, cotton or rag pulps automatically meet this criterion for no groundwood content.

The negative effect of acidity on permanence applies more to papers made with wood pulps than to papers made with cotton or rag pulps, but since most papers today are made from wood, the commonly used permanence standards require a final paper pH of 7.5 or greater. This century's increased air pollution levels, particularly the acidic conditions created by combustion gases from our automobiles and power plants burning fossil fuels, also contribute to the deterioration of paper. Thus permanence standards require papers to have a minimum 2% alkaline reserve of calcium carbonate or an equivalent buffering agent. This alkaline reserve buffers the paper, so that acidic atmospheric compounds will react with the alkaline filler before reacting with the cellulose fiber and the chemical bonds between fibers holding the paper together. Thus the paper will remain stable until the alkaline reserve is used up. Alkaline-sized papers (which typically use $CaCO_3$ for filler) will usually meet this requirement for an alkaline reserve. On the other hand, many papers are now being sold as "acid-free," which indicates only that an alkaline or neutral sizing process was used. Such a designation does not mean that a minimum pH of 7.5 has been achieved, or that a 2% alkaline reserve is present. Because these and other factors affect permanence, buyers should not equate this "acid-free" label with a guarantee of archival qualities.

The standard for paper permanence most widely used in the United States was developed in 1984 by the National Information Standards Organization (NISO), and this standard is published by the American National Standards Institute (ANSI).[177] There are recycled papers currently being made which meet both the ANSI and other permanence standards.

176

Pulp & Paper North American Factbook: 1990, p. 178.

177

This standard is known as the ANSI or ANSI/NISO standard. To meet it, paper must have the following characteristics:

Fiber Content: No groundwood or unbleached pulp

Minimum pH: 7.5

Minimum Alkaline Reserve: 2%

Fold Endurance: 30 (1 kg), MIT test method

Tear Resistance: 40 g for papers 74 g/m^2

The American Society for Testing and Materials (ASTM) has also developed paper permanence standards for some papers. The International Standards Organization (ISO) is currently developing international standards that, as proposed, are stricter than the ANSI standard.

178

Chandru Shahani, Library of Congress Preservation Research Office, interview, May 1991.

179

One reason for the care exercised in screening wastepaper for groundwood can be a mill's operating requirements. If a single-stage sodium hypochlorite bleaching system is used, the presence of groundwood fiber causes serious color reversion problems and darkening of fiber during deinking. Thus mills using this process have to be very careful. Groundwood content, in contrast, does not interfere with hydrogen peroxide bleaching, and mills using such a system have the capacity, at least, to tolerate groundwood content.

180

Not all research chemists agree that moist oven aging is an accurate predictor of permanence, or agree on the number of years to be equated with given conditions. This testing procedure is, however, the one method most widely used to evaluate paper aging characteristics.

To produce permanent papers, the objective for a mill producing recycled papers is essentially the same as that for a virgin mill— to eliminate the lignin content and produce paper in an alkaline environment to the proper strength specifications. Just as a virgin mill must be concerned with chemically processing and bleaching its pulp correctly, in order to remove lignin effectively, so must a mill using recycled fiber take care to control its wastepaper sources, to limit the inclusion of groundwood-containing papers in their fiber supply. As long as this is the case, and the paper is made in an alkaline environment to the proper specifications, recycled paper can be made which meets current permanence standards.[178] Many recycled mills are, in fact, very careful about screening wastepaper for groundwood. Incoming paper bales are sprayed with compounds, phloroglucinol or c-stain, that turn color in the presence of groundwood. If significant quantities of groundwood are detected, the wastepaper bales are rejected.[179]

The factors that contribute to paper permanence, and the causes of impermanence, are not completely understood. Research on the subject continues, and papers being made to current permanence standards have not been in existence long enough for us to know how they will, in fact, survive the centuries. Permanence is usually evaluated by the "moist oven aging" test, which is designed to simulate aging through an accelerated laboratory process. The paper sample is placed in an oven with controlled moisture conditions and an elevated constant temperature. Strength tests are performed on the paper before and after its stay in the oven. A certain number of days in the oven can be roughly equated to years, and this allows predictions to be made about how long the paper might be expected to last. These tests are carried out by a variety of laboratories, for research purposes and to test papers under consideration for specification. For any paper, virgin or recycled, this test is considered the most accurate and reliable predictor of longevity, and use of this procedure is recommended when permanence is a critical factor to paper selection.[180]

6 The Designer's Legacy: Closing the Loop

The public has embraced the idea of recycling as an intelligent and needed management approach to our solid waste problem, as well as to the conservation of resources. Collection programs are proliferating like crazy, widely adopted by citizens' groups and communities wishing to do their part to make the recycling circle work and hoping to alleviate the growing need to site new landfills or incinerators. But the success of recycling depends on the existence of markets for recycled products in order to sustain and build markets for the wastes themselves. In the 1980s and early 1990s the collection of paper wastes has frequently outpaced industry's capacity to use these materials. Thus it is the designers and their clients who specify the many tons of paper consumed, and who create printed materials that are more or less recyclable because of their design, who presently stand as the most influential players affecting this transition. The collective decisions of these individuals will, to a large extent, determine the success or failure of efforts to substantially increase recycling rates for paper in general, and for printing and writing papers in particular.

The majority of all printing and writing papers sold are used in offset printing. Specification of recycled papers has increased greatly in the late 1980s and early 1990s. (Photo © Michael Weymouth.)

The recent rapid increases in the collection of paper wastes have sometimes glutted the market with these potential resources, resulting in an oversupply that, lacking buyers, has been landfilled or incinerated despite having been carefully collected for recycling. This occurred in old newsprint markets in the latter 1980s, especially in the northeastern United States, and this scenario is becoming increasingly characteristic of the markets for office wastepaper. The collection programs for office wastepaper are expanding faster than is paper mill capacity to use them. Balancing the supply and demand for office wastepaper over the coming years will require substantial increases in the utilization of these wastes by manufacturers of printing and writing papers, tissue and containerboard mills, and other users.[181] Quite simply, the collection of wastes by itself does not make recycling work. Until new products are manufactured from these wastes and subsequently purchased, the loop is not closed.

The Paper Specifier's Impact

Before they make the multimillion dollar investments required to install deinking capacity, executives at printing and writing mills must be convinced that end users will buy recycled products and that they will gain a competitive edge in the marketplace by developing this capacity. If paper specifiers and buyers are able to create and sustain a strong enough market for recycled paper with a significant wastepaper content, the likelihood of positive investments will be greatly increased.

Government mandates at all levels have attempted to increase the markets for recycled paper. Despite the proliferation of these mandates, very few companies have made new investments for deinking capacity at printing and writing mills. The market for recycled products is not yet so strong that the evidence has compelled these paper mill executives to commit to major capital expansions for the use of wastepaper. In the long run, it may be the thousands of educated end users who ultimately exert the most influence on this issue— by creating viable markets for an increased number of recycled papers. The private sector purchases over twenty-three million tons of printing and writing papers each year, roughly 93% of the total U.S. consumption. The net effect of each private decision can thus contribute significantly to a collective effort that supports this transition. Each purchasing choice is significant. Some examples follow.

The Corporate Influence

Obviously, large corporations, with their tremendous purchasing volumes, are in a powerful position to influence the marketplace, though it is often individual consumers who are the grass-roots force behind the decisions of these companies to "go green."[182] Many companies have begun to insist on recycled papers for their catalogs, annual reports, and other materials. Each paper specification decision can have a dramatic impact. For example, the Patagonia Corporation prints several catalogs each year; most of the print runs are for quantities over 1.5 million copies. For the fall 1990

181
Franklin Associates, Ltd., *Supply and Recycling Demand for Office Waste Paper, 1990 to 1995: Executive Summary* (Washington, DC: National Office Paper Recycling Project, July 1991).

182
Tom Rattray, Associate Director of Corporate Packaging Development at Procter & Gamble, spoke frankly about the significance of consumer pressure on his company as the impetus for them to make substantial changes, increasing their use of recycled materials and reducing the amount of packaging materials used to sell their products. ("Packaging and Solid Waste— A Manufacturer's Perspective," presentation at *Pulp & Paper Wastepaper II Conference,* Chicago, IL, May 1991.)

183
Paul Tebbel and Ann Shilton, Patagonia Corporation, telephone interviews, February 1991.

184
Chapter 3 gives a more detailed analysis of estimates of tree consumption and solid waste generation per ton of paper consumed. The equivalents used here are 24 forty-foot trees consumed per ton of virgin paper produced and 3 cubic yards of landfill space required for each ton of paper discarded. If the paper specified is made with less than 100% recycled fiber, the savings are calculated proportionately.

185
Lory Christoforo and Lenore Lanier of the WGBH staff, with the help of their printers, diligently researched and tracked every single pound of paper specified by the Design Department for hundreds of different jobs over a six-month period. This figure was used to project the annual total. (Memos and interviews with Lory Christoforo, April and September 1991.)

catalog, the company purchased 610 tons of paper; for the spring 1991 catalog, it used 750 tons.[183] For both catalogs, the design staff selected printing paper with a minimum of 50% recycled fiber. Thus these two decisions alone, specifying a total of 1,360 tons of paper, resulted in the savings of an estimated 2,040 cubic yards of landfill space, equivalent to a volume of paper about 1.25 feet deep covering an area the size of an entire football field. Had the paper specified been made from 100% virgin fiber instead, approximately 16,320 medium-sized trees would have been required in addition, covering roughly 54 acres of forestland.[184]

The Design Group at WGBH Educational Foundation

The WGBH Design Department in Boston, Massachusetts, is a midsize design group that divides its work between the design of print material and that of video graphics for broadcast over Channels 2 and 44 TV. This group of seven senior designers, under the direction of one design director and with the help of several junior designers and production assistants, specifies over 300 tons of paper annually. They produce hundreds of different jobs a year. Some are small jobs, such as the run of 250 invitations for a special station event, requiring only 20 pounds of paper. Others are much larger, such as the monthly magazine sent to 185,000 station members and 100,000 copies of a 32-page teachers' guide. WGBH has increasingly made a commitment to using recycled papers, now choosing these grades for about 80% of all paper specified, including the printing of the monthly magazine.[185]

The private sector purchases 93% of all printing and writing papers sold in the United States. Combined federal, state, and local government purchases account for the remaining 7%. (Photo © Stanley Rowin.)

The Designer's Legacy: Closing the Loop

186
Paper with a 100% recycled fiber content has much more limited availability than does paper with a minimum of 50% recycled content. Some papermakers feel that there may be structural advantages to making papers with a mixture of recycled and virgin paper, rather than with 100% recycled fiber.

Claudia Thompson Design

My own office is staffed by myself and a full-time assistant. Much of our work is for small nonprofit clients and consists of modest, short-run jobs. Some of our work— architectural signage and other three-dimensional environmental graphics— is not print material at all, so relatively little paper is consumed in its production. Yet the annual total of paper specified from my office is over 15 tons. One job alone, a 40-page annual report for a local college, with a finished trim of 8.5 x 11" and a printing of 17,000 copies, required just over four-and-a-half tons of paper. Thus, during one year, by specifying paper made with 50 or 100% recycled fibers instead of virgin papers, I can save between 22.5 and 45 cubic yards of landfill space (equivalent to the volume of one or two modestly sized bedrooms) and eliminate the need to cut down about 180–360 forty-foot trees. I now specify recycled papers for the majority of the jobs that I design. Some is made with 100% recycled fiber, and much of it is manufactured with a mixture of virgin and recycled fiber. In specifying papers that have a minimum 50% recycled fiber content, I consistently seek out papers that meet this minimum through the manufacturer's purchase of postmill wastes.[186]

Calculating Your Tonnage

Calculating the tonnage of paper consumed for any given design job is simple, requiring only three pieces of information:

- *The total number of cartons of paper ordered for the job— information that can be easily obtained by knowing the quantities to be printed and the press configuration. Don't forget to include the amount of paper needed for the makeready, and be sure you are basing your calculations on the actual sheet size ordered for the job. Consult with your printer to verify the number of cartons ordered.*
- *The number of sheets of paper in a carton— information that is printed on the swatchbook.*
- *The M weight (weight per thousand sheets) for the sheet size of the paper ordered— this is also published on the swatchbook.*

The equation for determining total tonnage is:

$$\text{\# cartons of paper} \times \frac{\text{\# sheets}}{\text{carton}} \times \frac{\text{M weight (lbs.)}}{\text{1000 sheets}} = \text{\# lbs. of paper}$$

To convert to tons, divide the # of pounds of paper consumed by 2,000.

Appendix 5 has a complete form that can be reproduced and filled out to track paper tonnage specified or purchased. *Reducing the total amount of paper we consume unnecessarily must be our first strategy for dealing with the growing mounds of garbage we are creating.* For the paper we do consume, every ton of recycled paper used decreases the strain on our environment. It is instructive to take a look at the total amounts of paper you are specifying or purchasing on an annual basis and to consider the significance of your own decisions. It is gratifying to many of us to do our part by putting a few pounds of paper per day or week into a collection box. It is even more gratifying, however, as a paper specifier, to exert much more influence on the recycling loop through the power of a single decision.

Creating a Recyclable Product

The designer's legacy also includes the creation of a product that, after having finished its useful life, is more or less recyclable because of its design. Recyclability is a very relative term. It depends a great deal on what type of manufacturer will use the wastepaper, their capabilities, and the product to be made, as well as on what collection systems exist for recovery in the first place. The types of papers accepted in different collection programs will vary. Most printing and writing papers are made from bleached chemical pulp, making them potentially one of the best sources of secondary fiber available. Some printing and finishing processes, however, can quickly transform this premium fiber into a product that may be useless to virtually all the manufacturers who use wastepaper. To a deinking plant operator, anything other than the reusable cellulose fiber is considered a contaminant. An awareness of the effects of graphic arts processes on the suitability of the paper for recovery will help you create products that are, at least potentially, recyclable.[187]

Ink Coverage and Color Choices
Lighter ink coverage is, naturally, easier to deink than heavy ink coverage. Printers have been selling their printing trim wastes to wastepaper dealers for some time, however, and because they often overprint on stock during the makeready, some of their wastes are quite heavily printed. Generally a deinking plant can use a balance of heavily and lightly printed stock without encountering difficulties. Also, if the stock being printed on is cover weight or fairly substantial, the proportion of ink to fiber is actually

187
The final proof is in the collection and subsequent manufacturing into new products. Some states are now using the existence of operating collection programs with markets for wastes in their definition of the term "recyclable."

relatively small, even if solid ink coverage has been applied. For printed pieces, one could make the argument that light ink coverage is preferable, and certainly we do not need to design every piece with gobs of ink. However, in consideration of the human need for art and aesthetically pleasing imagery to stimulate the imagination, appropriate use of ink and color is desirable. A world without bright and cheerful solids and full-page halftones would be a little dreary. In sum, it is best to strive for balance. Occasional use of heavy ink coverage will not adversely affect the deinking process.

Some printing inks contain heavy metals. These compounds are largely associated with the pigments used to create certain colors, though they are also found in additives such as drying agents. Copper, for example, is widely used to make many blue and green pigments; barium is often used for some reds and yellows. Any heavy metals in the inks, once applied to paper, have the potential to contaminate our air or water regardless of whether paper is landfilled, incinerated, or recycled. Heavy metals in inks *do not* adversely affect the ability of paper to be deinked, but there is a concern about their potential danger that is completely independent of the recycling issue.[188] Currently, a variety of legislation is being considered by Congress that could restrict the use of certain heavy metals, potentially requiring some pigments to be reformulated without them. This has been controversial, since a few colors cannot presently be duplicated without these metals. The printing ink industry is engaged in research to find alternatives to some of the pigments currently in use. Meanwhile, designers may choose to limit their use of some colors as much as possible, and some references have listed the PMS colors most likely to contain various heavy metals.[189] Ink formulations may vary somewhat with different suppliers, however, so it is also necessary to make inquiries of the particular manufacturer whose ink you are buying. Unfortunately, getting accurate information about precisely how specific inks are formulated by a given supplier, and which inks contain exactly which heavy metals, can be quite difficult.

UV Coatings and Laminates

Ultraviolet cured inks and coatings are so problematic to most deinking and recycling mills that almost no purchaser of paper stock will knowingly buy wastepaper containing these materials; printers and graphic arts finishing houses that apply these coatings are rarely able to sell such wastes.[190]

188
See the section *Deinking Sludges*, in chapter 3, for a discussion of the naive argument that wastepaper should not be deinked because of heavy metals in ink.

189
A more detailed discussion of the heavy metals in inks appears in a special insert by Tedi Bish and Suzette Sherman, "Design to Save the World," *International Design*, Vol. 37, No. 6 (November/December 1990), pp. 49–64.

190
Mike Hecht, "High Volume Coating and Drying," *High Volume Printing* (August/September 1990), pp. 92–96; other detailed reports by paperstock dealers on the salability of wastepaper at several national recycling and wastepaper conferences, April and May 1991.

191

At least one deinking mill, the Memphis, Tennessee, plant of Ponderosa Fibres, has been successful in de-polying laminates such as milk carton stock and window envelopes. (Glen Tracy, Vice-President Ponderosa Fibres of America, telephone interview, March 1991.) This capability tends to be the exception rather than the rule, however.

192

Window envelopes with a glassine patch have sometimes been considered preferable to window envelopes with a plastic patch since glassine is actually a type of paper. Because it is typically waxed, lacquered, or otherwise treated when it is made, however, the glassine is not as easily recycled as the envelope paper itself. In addition, collection programs may be set up to disallow window envelopes with any sort of patch in order to avoid the risk of plastic laminates being inadvertently mixed in.

193

Richard W. Wand, "Paper Recycling's Contaminant Barrier," *TAPPI Proceedings*, Madison, WI (April 1989).

Thus a decision to use UV inks or super-glossy UV coatings renders the product essentially nonrecyclable. Laminates of plastic and paper also routinely cause contaminant problems to many manufacturers using wastepaper. Laminate products, such as plastic window envelopes, are often excluded from office collection programs, though in rare cases they may be allowed.[191] In order to create a product that can be readily recycled, none of these processes should be specified. If window envelopes are required, they can be purchased without a patch over the window at all. These "open-face window envelopes" are good enough for many jobs, and without a plastic film attached to the window, they are potentially more recyclable.[192]

Aqueous Coatings

Aqueous coatings have been touted as a recyclable alternative to UV and radiation cured coatings. In place of the separate plastic or film laminates, aqueous coatings exist first as a liquid solution, which is applied over the entire surface of the printing sheet by a unit of the printing press. The coating may be applied on-line, by one of the units of the press during the original printing process, or off-line, in a separate pass through a press after the inks have been printed. The coating solution dries quickly, giving the printed piece a very glossy appearance. Because these coatings contain polymers and other compounds designed to create a smooth, hard surface bonded to the fiber of the paper, they are quite tough if properly cured. Despite the claims of recyclability, I have been unable to locate any conclusive studies demonstrating that these materials are easier to process through a deinking plant than is paper to which plastic laminates or UV coatings have been applied. Certainly, the presence of *any* coating makes it more difficult to recover the fiber.

Bindings and Glues

Water-soluble glues and adhesives *are* readily recyclable, since they will dissolve during the repulping and deinking process. However, hot-melt glues, which are commonly used in the perfect binding of books and the construction of folders, boxes, and other items, pose major contaminant problems to secondary fiber mills.[193] These glues tend to glob together as small particles in the pulp, showing up later on the paper machine, often creating shiny spots or holes in the paper. In addition to causing manufacturing flaws, these "stickies" build up on the wire screen of the paper machine, requiring extra cleaning and care of equipment, including the use of solvent-based washes to break them down. They can also cause major breaks to occur during the manufacturing process, a dramatic event that is quite unwelcome at the mill, with paper traveling through many of the machines at speeds over 35 miles per hour! Consult with your printer, bindery, or finishing house to specify water-soluble adhesives wherever possible.

Pressure-Sensitive Adhesives

Many of the pressure-sensitive adhesives on the market, used for self-seal envelopes, tapes, pressure-sensitive labels, the ubiquitous office stick-on notes, and other products, are also non-water-soluble and difficult to recycle. Like the hot-melt glues, these adhesives tend to be a major source of stickies.[194] While a few mills (usually those making tissue or boxboard) will buy wastepaper containing these materials, most will not. This problem is starting to be addressed by manufacturers, some of whom have introduced label stock and other products made with more soluble adhesives, which they advertise as "repulpable" or "recyclable." However, it will be impossible for most collection programs to differentiate between acceptable and nonacceptable stick-on notes, labels, envelopes, and so forth. Thus collection rules will tend to be written to avoid them altogether, and wastepaper dealers will not buy paper containing these adhesives unless they have a specific purchaser able to handle the most problematic of these contaminants. Until the problem is addressed industry-wide from a manufacturing standpoint, and as long as many of the pressure-sensitive products on the marketplace are a source of stickies, these products will, practically speaking, be nonrecyclable.

Foil Stamping

The reflective and transparent foils that can be applied to paper by a stamping process use a combination of heat and pressure to activate the adhesive on the back of the foil. This also creates a nonrecyclable product, because the foil is firmly bonded to the fiber and almost impossible to remove from the pulp. Designers who specify this finishing technique should consider the life expectancy of the product and its potential to be recycled by the end user before choosing a process that effectively sentences the paper fiber to a single life.

194
Ibid., pp. 4–5.

Thermography

This process is used by printing houses to create raised letters on paper, presumably simulating older engraving techniques. It requires printing with special inks that, after being dusted with resinous powder, are heated so that the printed area rises from the surface of the paper and fuses into a hard surface. Because of the resins used and the heat-set nature of the process, it is again very difficult to extract the cellulose fiber from paper printed with this technique, and such wastepaper is likely to introduce difficult-to-handle contaminants into most deinking systems. Thus the use of this technique also creates an essentially nonrecyclable product.

Varnishes

There is lack of agreement about the extent to which the use of ink varnishes affects the recyclability of paper. Some deinking plant personnel have stated that they do not anticipate difficulty in using wastepaper with occasional varnishes. Confirming this, the printers I interviewed do not report difficulty in selling their wastepaper with both inks and varnishes applied. They may, however, get a slightly lower price for their wastepaper if varnished wastes are included.[195] Trying to make a generalization about the recyclability of varnishes is difficult, since it is clear that such wastes have a market and are being used, but it is possible that the formulation of varnishes (somewhat different from inks) might pose a problem to some mills.[196] Any such difficulties would depend on the percentage of these materials in the wastepaper and the sophistication of the deinking operation. Based on current information, the lack of definitive studies one way or the other, and the fact that some deinking plants are using these wastes, I would conclude that the use of varnishes does not generally pose a problem. Clearly their use is far preferable to the use of plastic laminates or UV coatings.

Embossing and Die-Cutting

Fortunately, there are a few graphic arts techniques available that allow for a wide variety of creative possibilities and are completely benign to product recyclability. Unlike even the application of normal printing inks, embossing and die-cutting add no contaminants to the fiber but merely alter the surface of the paper. For the designer with a comfortable production budget and an environmental consciousness, there is nothing better than embossing or die-cutting. After writing a long list of cautions, it's great to say, "Do it!"

195
Adam Daniels, Daniels Printing, telephone interview, June 1991.

196
Tom Garbutt, CRS Sirrine Engineers, Pulp & Paper Division, interview, May 1991.

Soybean Inks

Other environmental factors are increasingly being considered by designers, clients, printers, and publishers. While we cannot cover all of these issues, the current interest in and confusion about soybean inks does deserve attention here. Like the term "recycled paper," the term "soybean ink" includes a very broad range of products for which there are no binding or uniform standards.

All inks are made with three principal components— pigments, solvent vehicles in which the pigments are dissolved, and additives.[197] In most printing inks the vehicles are oil-based, and petroleum-based oils are the type most commonly used.[198] In vegetable-based inks, some portion of the petroleum oil has been replaced by vegetable oil. Soybean oil is the vegetable oil most commonly substituted, but corn oil, cottonseed oil, linseed oil, and others are used as well. The substitution percentage can vary greatly. Sometimes only very small amounts of the petroleum oil are replaced; more commonly, about 50% of the petroleum oil may be substituted. Only in rare instances is 100% of the petroleum oil replaced by vegetable oil. Variations will occur with different ink manufacturers and with the particular formulation of different types of inks, and no fixed percentage is consistently used.

The American Soybean Association (ASA) has licensed two versions of a trademark known as the "SoySeal." The version licensed to printers and publishers includes the words "Printed with Soy Ink," while the version licensed to ink manufacturers includes the words "Contains Soyoil." Neither licensing agreement requires the ink to meet any minimum content standards for the total amount of soybean oil in the ink. For both trademarks, the licensing agreements only require that the seal be used with products containing *some* vegetable oil, and that the vegetable oil used be exclusively soybean oil except for trace amounts.[199] In late 1990 the National Association of Printing Ink Manufacturers (NAPIM) recommended to its members that minimum target percentages of soybean oil be met for the formulation of so-called "soy inks." For sheetfed printing inks the recommended minimum percentage was 20% of total ink ingredients, although it was not specified whether this percentage was by weight or volume.[200] Ink manufacturers have typically interpreted this recommendation in terms of ink weight. To meet such a requirement, sheetfed inks might be formulated with an oil content that is roughly one-half to two-thirds soybean oil and the rest petroleum oil.[201]

197
National Association of Printing Ink Manufacturers, *Printing Ink Handbook* (Harrison, NY: 1988).

198
While most inks are oil-based, some flexographic inks are water-based. These inks have proven to be problematic in some deinking systems, although this is not universal. (Discussion at *New England TAPPI/Connecticut Valley Pima Technical Seminar on Deinking,* March 1991.)

199
American Soybean Association, "Memorandum of Understanding: Soyoil Certified User Agreement" and "Memorandum of Understanding: SoySeal User Agreement"; Stu Ellis, American Soybean Association, telephone interviews, June and September 1991.

200
National Association of Printing Ink Manufacturers Bulletin, "Soybean Oil Content of Printing Inks," October 15, 1990. For heatset inks the minimum recommended content of soy oil was 18%; for forms ink it was 40%, and for newsprint ink it was 55%. Based on its 1990 surveys of ink manufacturers, the American Soybean Association is now recommending that targets for heatset inks be lowered to 10%. The ASA has also since stated that its recommended percentages are to be based only on the "nonpigment portion of the ink." Again, it has not been clarified whether this percentage is by weight or volume. (Stu Ellis, American Soybean Association, telephone interview, September 1991; American Soybean Association brochure, "Put on a White Hat.")

201
Ralph Viscione, Century Ink Co., telephone interviews, June and September 1991.

202
Pam Loman, "Making the Decision to Go Soy," *High Volume Printing* (February 1991).

203
Daniels Printing in Everett, Massachusetts, ran these controlled comparison tests for soy versus petroleum inks, including printing large areas of solids and four color halftones. I reviewed their printing samples and honestly could see very little difference between them. The company did decide to adjust many of their presses in order to run soy-based inks routinely.

The American Soybean Association licenses these trademarks for use by printers and ink manufacturers. There are no minimum soybean oil content percentages that must be met for use of these symbols. (Reproduced by permission.)

The most immediate advantages from using vegetable or soy inks are felt by the printer. If made with a high enough concentration of vegetable oil, these inks release far fewer volatile organic compounds (VOCs) than do petroleum-based inks. This makes the printing plant environment much safer, more pleasant, and more readily in compliance with the requirements of the reauthorized Clean Air Act (1990). In addition to aiding printers, the lower VOC emissions of vegetable-based inks help all of us by reducing the release of air-polluting hydrocarbons into the atmosphere. However, the pigments in these inks, and the drying compound additives, are not changed by the substitution of vegetable for petroleum oil; they may still contain heavy metals. Thus soy inks, by themselves, are not a solution to all the environmental concerns associated with inks.[202]

The print quality achievable with soybean oil inks is excellent. Some printing personnel have even expressed a preference for the printing characteristics of soy-based inks, commenting that they give better ink hold-out and color rendition, especially on uncoated papers. In a controlled test conducted by one printer on coated papers, comparing soy inks (in which roughly 50–75% of the petroleum oil had been replaced) to purely petroleum-based inks, very little perceivable difference was evident. Press-operators, looking closely at photographic reproductions, felt that the halftone dots printed a bit more cleanly with the soy inks.[203] The one drawback is that soy inks may require a bit more drying time than conventional inks due to their lower volatility; therefore, printers must sometimes allow additional time between runs before printing the second side of a paper. Soy inks that are specifically formulated to run on uncoated paper can have a higher soy oil content than inks used for coated papers, due to the increased absorbency of uncoated papers. On coated paper, because of the longer drying time required by soybean oil and the impervious nature of the paper, there may be some limitation on the total amount of petroleum oil that can be replaced.

7   Choices and Opportunities

This decade will be a critical juncture for national policy in the arena of waste management and recycling. The availability of landfill space continues to decrease significantly, while the total amounts of waste generated in this country are still rising. The pressure to build new incinerators will mount as more and more landfills are filled to capacity and closed. Fights over the siting of new landfills and incinerators will remain contentious, and with good reason. No community wants to bear the environmental consequences of being a dumping ground for tremendous amounts of garbage generated regionally by thousands of other communities. The solution to our solid waste problem requires all of us to consume less in the first place, to reuse materials as long as their useful life permits, and to recycle as much as we can. As the largest component of the MSW stream, paper is of particular importance.

Compared to all other industrialized areas of the world, wastepaper utilization in the United States is relatively low. Substantial overall increases in the utilization rate will require increased use of wastepaper to manufacture printing and writing papers— above the 6% rate of 1990. This in turn will require large investments in deinking capacity at mills that make these papers, to supplement the expansion of wastepaper pulping capacity already underway in the newsprint, tissue, and paperboard industries. The likelihood of this outcome will be furthered by the development of meaningful national standards for recycled paper that honestly distinguish the products that derive a minimum recycled content from postmill wastes. Without such standards, consumers are vulnerable to deception and trapped in a marketplace wherein significant choices between competing products are obscured.

(Illustration © Judy Love.)

Comparisons with Other Countries

In 1989 the per capita consumption of paper and paperboard products in the United States stood at 670 pounds, while the per capita consumption averaged only 336 pounds in the EEC countries of Europe and 206 pounds in all European countries. Paper consumption averaged 491 pounds per person in Japan. By contrast, in the former Soviet Union it was 79 pounds, and in China it was only 28 pounds.[204] Not only do Americans consume more paper, on both a per capita and a total basis, than any other country in the world, but generally speaking we rely much more on virgin fiber to manufacture it. A comparison of wastepaper utilization rates around the globe shows the United States to be far behind most other countries in its use of wastepaper as a fiber source for new paper products.

In 1989 the U.S. wastepaper utilization rate stood at 26%, while in that same year wastepaper contributed about 34% of the fiber used throughout Europe, with some countries reaching remarkably high levels. Denmark and the Netherlands led the continent with 68% and 66% utilization rates, respectively. Spain and the United Kingdom followed at roughly 60%. France, Germany, Italy, and Switzerland all had utilization rates between 45 and 50%.[205] On the other side of the globe, Japan used recycled fiber for over 50% of its paper production in 1989, while the rest of Asia, excluding China, used it for almost 70%.[206] On our continent, Canada uses an even higher percentage of virgin fiber to supply its mills than does

204
Pulp & Paper North American Factbook: 1990, pp. 366–367.

205
European data from "European Wastepaper Demand to Increase 50% by 2000," *Paper Recycler,* Vol. 1, No. 2 (November 1990), p. 7. Only the forest-rich Scandinavian countries utilize very little recycled fiber.

206
John A. Latham, Jaakko Pöyry Consulting, "Recycled Fiber: Opportunities for Mills and Suppliers," paper presented at *New England TAPPI/Connecticut Valley Pima Technical Seminar,* Holyoke, MA (March 1991).

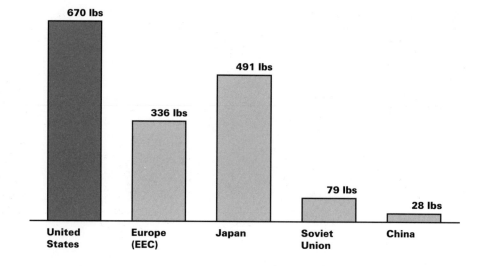

Figure 7.1
Per Capita Paper Consumption: 1989

Residents of the United States use more paper and paper products than those in other regions of the world. Figures show the average total consumption per person during 1989. (Data from Pulp & Paper North American Factbook: 1990.*)*

the United States, and the 1988 North American wastepaper utilization rate was only 23%. This utilization rate is "the lowest of any major geographic region in the world, even though North America is the world's largest producer of paper and paperboard."[207] With such low wastepaper utilization rates, with the world's highest paper consumption rates, and with paper as the largest component of solid waste, it is not surprising that we are in the midst of a mounting solid waste crisis.

[207] Ibid., p. 1.

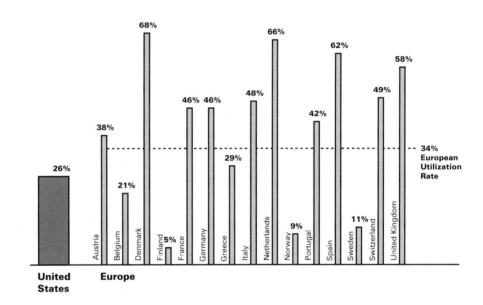

Figure 7.2
Wastepaper Utilization Rates in the U.S. and Europe Compared: 1989

While wastepaper utilization in the United States stood at 26%, wastepaper provided about 34% of all fiber used for papermaking on the European continent. Wastepaper utilization rates for the European countries vary as shown. (Data from Paper Recycler and American Paper Institute.)

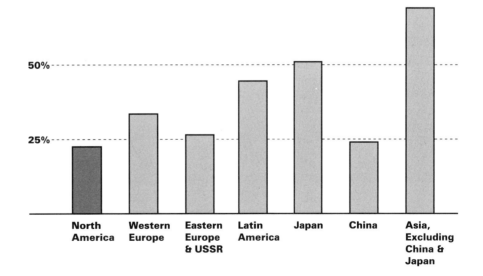

Figure 7.3
Wastepaper Utilization Rates Around the Globe: 1989

North American paper mills utilize wastepaper at a lower rate than do paper mills in any other major geographic region on earth. (Data from Jaakko Pöyry Consulting.)

[208]
Franklin Associates, Ltd., *Characterization of Municipal Solid Waste in the United States: 1990 Update*, p. 46.

What You Can Do: A Waste Management Hierarchy

A drastic change in our habits is required to address the environmental consequences associated with these behaviors. We must establish a waste management hierarchy with a logical sequence of steps. Both the data and common sense lead directly to the conclusion that our very first priority must be to limit our consumption of goods in order to reduce the total amount of garbage we generate. In waste management lingo, this is called "source reduction."

Source Reduction

There are many opportunities to reduce our overall consumption of printing and writing paper, and each of them should be explored. The savings can be significant, in environmental terms as well as in dollar amounts:

- *Before initiating production of a new printed communication piece, question whether the piece will effectively serve its purpose and whether it is worthy of creation to begin with. Will the piece produced justify the paper consumed and waste generated? Is it a product, event, or service worth promoting? Perhaps there is a better way to communicate the message, from an environmental point of view as well as in terms of total effectiveness. Perhaps the piece simply doesn't deserve to be done.*

- *Once a project is initiated, examine ways to reduce the total amount of paper consumed throughout the design process. Reduce the size of the piece wherever possible. Making it smaller will not necessarily compromise the design; sometimes the limitation may make it more effective.*

- *Design each piece with dimensions and a format that maximize the use of the press sheet and minimize the trim.*

- *Use the lightest basis weight stock possible for the job. For example, a long run on 70 or 80 lb. text instead of 100 lb. text saves a lot of resources and generates significantly less waste.*

- *Make it a standard practice to use both sides of letterhead papers for correspondence, and both sides of the sheet for photocopies and laser printing. This simple behavior change will lead to a substantial savings of paper.*

- *Take reusable cloth bags to carry goods purchased while shopping. This will result in a tremendous savings of paper (and plastic), since in 1988 we discarded 2.7 million tons of paper bags and sacks in the United States.*[208]

Reuse

The second logical rung on the hierarchy of waste management is reuse, which offers two major benefits. First, resources and energy are conserved, and the pollution associated with manufacturing is avoided, since fewer new products are needed to replace existing ones. Second, we extend the life of our landfills and reduce the pollution associated with landfilling and incineration, since we are disposing of less. In the past, reuse was more common in the routine of daily life, and it now needs to be revived. Products need to be designed functionally, to serve as long a life as possible, rather than being disposable. Reuse is a strategy that applies particularly well to durable goods such as cameras, razors, and many other consumer items, but paper can be reused as well, in many creative ways:

- *In the office or at home, paper printed on one side can be reused for scrap paper, rough drafts, sketches, telephone messages, and the like. Don't discard paper until it has been completely used.*

- *Reuse office products such as envelopes, cardboard inserts, boxes, and other packaging. Use button envelopes, which are designed to be circulated many times before being discarded.*

- *Outdated and unneeded swatchbooks and paper promotions can be donated to schools and day care centers, which often welcome having attractive colored papers to use for student projects.*

- *Many papers can be reused for packaging. Old posters, comic pages, and other papers can make wonderful gift wrap. Paper bags can be reused to wrap packages for mailing or other purposes.*

- *Commercially, some papers are being collected and processed for reuse (as opposed to recycling) in new products. For example, old newsprint is being shredded and processed for animal bedding or cat litter as well as molded or other protective packaging materials. Because this reuse option limits the number of times the paper can subsequently be used, however, some waste management officials are concerned that these uses not replace the recycling of these paper wastes into new paper products, for which the loop can be traveled many more times.*

209

There has been some confusion about the recovery rates for paper in comparison to other materials. In 1990 the paper industry widely promoted the pie chart shown here under the heading "Paper Leads Recycling," stating that paper products represent 86% of all materials recovered for recycling. This statistic was generated by comparing the weight of paper materials collected from MSW to the weight of aluminum materials collected, plastic materials, glass, and so on. It does not reflect the amount of each material originally generated, and therefore is not in any way a comparison of the recovery rates for these different materials. (Data from American Paper Institute press kit, February 1990.)

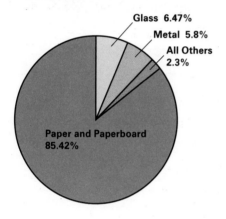

210

Franklin Associates, Ltd., *Characterization of Municipal Solid Waste in the United States: 1990 Update*, pp. 45, 78.

Recycling: Making It Work

Even if we reduce the total amount of paper we consume and reuse every piece as long as we can, it is clear that large amounts of paper will continue to be consumed. Thus recycling is the final crucial step in an integrated waste management hierarchy. In order to make a real dent in our solid waste problems we need to increase recycling rates beyond the historical precedents established during earlier times, especially the world wars.

Not every material in the solid waste stream is recovered for recycling equally, and while 13% of MSW was collected for recycling in 1988, 26% of all paper products were recovered.[209] A study for the EPA projects that, by 1995, the average recovery rate for all MSW materials will reach 20–28%. This projection assumes that recovery rates for most materials will increase, with 31–38% of postconsumer paper wastes recovered by 1995. If present trends of increased consumption continue as projected, however, even reaching these targeted recovery rates will leave about the same quantities of wastes requiring disposal.[210] Some waste management officials, environmental advocates, and members of the paper industry have started to ask whether we can achieve recycling rates for paper that are even higher, so that real progress can be made toward solving the solid waste issue. Many believe that with a concerted effort by all participants— design professionals, corporate print buyers, citizens, industry executives, and government officials— we can. To help further this objective:

- *Specify recycled papers for the majority of jobs you produce. There are recycled papers appropriate to almost every design need.*

- *Select recycled papers with high percentages of postconsumer, deinked, and postmill wastes, to support markets for the wastepaper already being collected. In the absence of a national standard for recycled content, this requires asking detailed questions of mills and suppliers and using the guidance provided by existing certification programs such as the New York State labeling program and the Canadian EcoLogo program.*

- *Buy recycled paper for copiers, laser printers, and other office uses.*

- *Buy other paper goods made with recycled materials— tissue products and paperboard packaging. (In the consumer market for toilet tissue, paper towels, and so forth, many of the economy or private label tissue products use recycled fiber for some of their furnish, even if they are not labeled as recycled. Typically, the highest-priced "premium" tissue products are made exclusively with virgin fiber.)*

> • *Stay educated on the national debate in this country about definitions and standards. Make your opinions known to the EPA, the FTC, and other bodies involved in the development of definitions, procurement guidelines, and labeling requirements. In order to be effective, any standards developed must be acceptable to the majority of the end users who purchase paper.*

The importance of buying recycled products cannot be overemphasized. The paper recovery rates reported here will not equal recycling rates, if markets for the wastes do not exist and the collected wastes end up being landfilled or incinerated.

Future Decisions

Given the commitment of the U.S. paper industry to increasing wastepaper utilization, primarily in paperboard, newsprint, and tissue production, the country is expected to achieve an industry-wide utilization rate of 30% by 1995. On a worldwide basis, this is still a fairly low target, and it is lower than the 37% utilization rate that the country achieved during World War II. If much of the rest of the world has achieved higher rates, and we have done so in the past, why can't we do so in this decade? The key to the problem rests with the printing and writing mills. Their products now represent such a significant portion of our total paper consumption that, without substantial increases in wastepaper utilization for these papers, we will soon approach the limits to what is attainable.

Yet, despite how quickly the public has embraced the idea of recycling, despite the tremendous number of papers being sold and labeled as recycled, why has the wastepaper utilization rate for printing and writing papers remained at 6–7%? Why are so few investments being made by these manufacturers? The answer is simple. The single biggest impediment at the moment is the lack of an enforceable and meaningful national standard for the marketing and sale of these papers. Since everything from paper made exclusively with mill wastes to paper made exclusively with postconsumer wastes recovered from the MSW stream is being sold as recycled, drastic differences between products are obscured. If buyers were more readily able to distinguish between competing products, change would be more rapid. The "free" in the term "free-market economy" is predicated on the existence of accurate information that allows consumers to make real choices. Clearly, a standard is needed.

An Effective National Standard

There is a simple way to ensure that a national standard for paper sold as "recycled" will be meaningful and effective in further developing markets for wastepaper, and that is to require the wastepaper that supplies the minimum recycled fiber content to come entirely from postmill wastes, purchased from independent business entities in which the paper manufacturer has no financial interest. Buyers would then be assured that when they buy recycled paper, that minimum percentage has been supplied through the collection infrastructure associated with recycling efforts.

Arguments have been made that definitions must accommodate those mills (primarily paperboard, newsprint, or tissue mills) that use 100% recycled fiber, and that in these cases mill wastes are genuinely from recycled sources and therefore constitute recycled material. These arguments cloud the issue, since mills using 100% recycled fiber will already be making products that exceed any established minimum content requirements. The central issue remains: *What are the minimum requirements that all papers of a given type should be expected to meet?* The precedents established in the printing and writing industry from 1988 to 1991 indicate that without such a postmill requirement, a large number of "recycled" papers will be made and sold for which the recycled content comes either entirely or primarily from paper mill wastes. Yet very few customers consider that to be the mark of a recycled product.

The purpose of establishing minimum standards is certainly not to set a ceiling on the total amount of wastepaper used in a product; on the contrary, it is reasonable to hope that products will sometimes not only meet but also exceed the minimum requirements. No rational person would argue that mill wastes should not be recycled, but it is appropriate that these wastes, which have always been recycled into new paper, be used to supply that portion of the recycled content that exceeds whatever minimum requirements are established. Convoluted definitions and standards, allowing only certain types of paper wastes within a mill or allowing a variable percentage of the mill broke itself, will be complicated for mill personnel to track and perhaps impossible to audit or verify. This also makes them much more likely to be abused. On the other hand, because most mills keep careful records of their wastepaper purchases, compliance with a postmill standard will be easier for the mills in terms of tracking their own manufacturing procedures, and it will be readily verifiable to an outside certification or inspection process.

Such a national standard, which clearly outlines substantive requirements for the manufacture of recycled paper, would also give the paper industry needed clarity in making the difficult decisions about future investments. With it, executives will be able to develop realistic design and cost estimates for installing the deinking equipment required to make paper that meets such a standard. Since the investments would be made for a long-term instead of a short-term manufacturing change, the costs versus benefits could be better analyzed: Will the investments required be offset by the competitive edge gained from the marketing of recycled papers that customers can buy with the assurance that the product meets a known and respected standard?

The stakes of environmental marketing strategies are high, since consumers have shown themselves quite willing to make purchasing choices based on environmental preferences. Debate over this issue is likely to continue until, eventually, a national standard emerges that has enough public consensus behind it for the issue to come to rest. The question is not so much will there be a standard, but what form will it take and by whom will it be developed and implemented?

In the resolution of these issues, it is our broader perspective that is most important, for the ultimate stakes are not marketing strategies but the stability of human life on planet earth. The pressures to solve the many environmental crises we face are mounting. We can no longer afford to overlook the infinite interconnections between the challenges before us and our common interest in solving this and other issues in a manner that recognizes the larger whole. Each of us has the power to participate constructively in the creation of a future that will sustain us, and it is through our individual actions and commitment that this can be accomplished.

Appendices

Appendix 1
Glossary

Appendix 2
Pulping & Papermaking Processes

Appendix 3
Bibliography & Resources for Further Information

Appendix 4
Recycled Papers Available

Appendix 5
Designer Impact Analysis Form

Appendix 1

Glossary

Basis weight:
The weight of a ream (usually 500 sheets) of paper cut to a specified size that is standardized for that type of paper. This can be confusing because the standard size for measuring basis weights is not the same for all types of paper. For example, the sheet size for measuring the basis weights of bond and writing papers is 17x22", while for text papers it is 25x38", and for cover stocks it is 20x26".

Brightness:
Paper brightness values are a measure of the reflectivity of paper to light under controlled conditions, using a single wavelength of light. Brightness values for paper range hypothetically from 0 to 100; most white offset papers have brightness values between 75 and 90. Papers with the highest brightness are made by adding fluorescent dyes to increase reflectivity. Several different test methods for measuring brightness are used; they may yield slightly different numbers for the same papers.

Calendering:
The process of pressing paper through rollers (usually of hardened, polished metal) to increase its smoothness and/or gloss. The calendering of machine-made papers, when performed, is done at the end of the manufacturing process. This same effect has been achieved in the production of handmade papers by a variety of means, especially hand-polishing with stones or other hard-surfaced materials.

Chemical pulp:
Pulp made through a variety of processes, all of which involve cooking the fibrous raw material (most commonly wood) with chemicals, in order to extract the cellulose fiber; the most common are the sulfate (kraft), sulfite, and soda processes.

Dandy roll:
A cylinder roll, covered with wire mesh, positioned in the final third of the wet end of the paper machine, which rotates over the wet web of the paper as it is being formed. Invented in the nineteenth century, the dandy roll is used to apply finishes (such as laid, wove, or linen) or watermarks to machine-made papers. Plain dandy rolls are also sometimes used in order to achieve good sheet formation.

Deinking:
Pronounced de-inking, this process removes applied inks, finishes, glues, and other contaminants from wastepaper in order to extract the cellulose fiber. Typically this requires extensive processing through a variety of pulping, screening, cleaning, washing, and/or flotation equipment.

Dioxins:
A large class of chlorinated organic compounds, known commonly as dioxins and furans, or sometimes simply dioxins. There are 75 polychlorinated dibenzo-p-dioxins (PCDD) and 135 polychlorinated dibenzofurans (PCDF) known to exist. Dioxins are proven to be potent carcinogens in animals other than humans and are widely suspected to cause cancer and birth defects in humans as well. They were a major contaminant at Love Canal, at Times Beach, Missouri, and were a key ingredient in Agent Orange, which was used extensively as a defoliant during the Vietnam War.

Dot gain:
The enlargement of the size of halftone dots in the reproduction of photographs through offset printing. Some increase in the size of the dot may take place in the transfer of image from film negative to printing plate, and especially in the final transfer of ink from printing plate to paper.

Fiber-added papers:
Papers to which fibers have been added primarily for aesthetic reasons. There are many papers, both virgin and recycled, in which material such as fine colored threads or wood shives are added to give the paper a textural feeling. Sometimes punched-out pieces of paper known as planchettes are added for their aesthetic effect or, in the case of bank note papers, to make counterfeiting more difficult.

Fibrillation:
In papermaking, the process of raising fibrils or little tiny threadlike hairs from the surface of the whole cellulose fiber so that bonds will form between the fibers. This is done during the beating of the fiber, and in most machine-made papers it is now a function of the refining process.

Formation:
The distribution of the physical components of paper, most notably the fibers, throughout the sheet. A uniform distribution is a mark of good sheet formation; this contributes to superior paper properties and ease of printing. If the fibers are distributed unevenly, this gives a lumpy sheet formation, and the resulting variability in the thickness of the sheet will make printing more difficult. Sheet formation may be observed by holding the paper up to light to see how evenly light is transmitted through it.

Free-sheet:
Paper made with chemical pulps that have been processed and bleached to remove lignin and other extraneous compounds. As of 1990, industry standards allowed free-sheet papers to contain a maximum of 10% groundwood or mechanical pulp. These pulps may also be called "wood-free," to indicate that they have been processed to the point that they consist primarily of the cellulose fiber portion of the wood.

Furnish:
The materials used to supply the various components of the liquid stock from which paper is made. The principal ingredient is fibrous pulp material made from either processed virgin plant material, recycled wastepaper, or a blend of the two. Other ingredients include fillers, sizing agents, and other additives.

Groundwood pulp (also called mechanical pulp):
Wood pulp produced by mechanically grinding the wood into short fibers. The groundwood process was the first method developed to pulp wood; it is still commonly used to make newsprint but much less frequently used to make printing and writing papers. Groundwood pulp is cheaper to manufacture than chemical pulp and yields the most paper per ton of wood, but this process does not remove the lignin or other extraneous compounds from the cellulose fiber.

Lignin:
Binding compounds within the cell structure of vascular plants, especially trees, which if not removed from wood pulp make paper subject to rapid deterioration. Lignin is not removed by mechanical (groundwood) pulping processes but can be removed through chemical pulping processes.

MSW (municipal solid waste):
Wastes collected by municipalities from residential, office, commercial, institutional, and a limited number of industrial sources. Municipal wastes do not include many industrial process wastes, municipal sludges, or incinerator ash, even though these wastes sometimes do end up in municipal landfills or incinerators.

Makeready:
The period of time at the beginning of the print run during which the necessary press adjustments are made to bring the print quality up to desired standards. The makeready typically consumes significant quantities of paper, which is often turned over and reused before being discarded. For any job the paper buyer must order enough paper for both the final quantity of printed pieces and the makeready.

Mechanical pulp:
See groundwood pulp.

OCC (old corrugated containers):
A large category of wastepaper consisting primarily of old corrugated containers and cuttings from the manufacture of containerboard, paperboard, and kraft grocery bags.

ONP (old newsprint):
A wastepaper category including mostly old newspapers but also related wastes such as over-issue news and printed or unprinted scrap from newspaper printing plants.

Opacifier:
A substance added to the paper during manufacturing to make it more opaque (less translucent), such as clay, calcium carbonate, and titanium dioxide.

Perfect binding:
A bookbinding technique in which the folded signatures of paper are trimmed and glued as individual sheets into a paper cover. Perfect-bound books are more likely to have their pages fall out than are books bound with sewn signatures. Because of relative costs, however, most paperback books are perfect-bound.

Preconsumer waste:
Waste generated before a product has reached its final end user. For paper production this includes a broad category of both unprinted and printed paper scrap generated by paper mills, converters, printers, publishers, manufacturers, and a variety of other businesses. Only some of this waste requires deinking.

Postconsumer waste:
As defined in the EPA Guideline, this consists only of paper wastes collected from offices, retail stores, and residences "after they have passed through their end usage as a consumer item." Upon careful questioning about wastepaper sources, however, it is evident that some mills and wastepaper dealers are taking liberties with their classification of postconsumer materials. In addition, some organizations use a definition that differs from the EPA definition and includes a broader range of wastepaper.

Postmill wastes:
Wastepaper recovered from sources after the original manufacturing, trimming, sheeting, and converting operations are completed. Not all users of the term agree, but many feel that in order to be considered postmill, these wastes should be purchased from independent businesses *in which the manufacturer has no financial interest.*

RCRA (Resource Conservation and Recovery Act):
An act passed by the U.S. Congress in 1976. Among other requirements, RCRA mandates the EPA to institute guidelines for federal purchasing that promote the use of paper and other products made with recovered materials.

Refining:
The process of cutting and roughening cellulose fibers during papermaking to control fiber length and bonding characteristics. Depending on the type of papers being made and the original fiber source, pulps will be refined quite differently.

Secondary materials and secondary waste:
Terms used by the states of New York and California, respectively, to define wastepaper that qualifies for providing recycled content. In both cases, all such wastes must come from sources *following* the paper manufacturing operation. No sawdust, forest residues, dry or wet mill broke, or wastes generated within a paper mill or within the same parent company qualify.

Sizing:
Material added to paper during manufacturing, either internally as part of the paper furnish, or externally as surface sizing, to make it more impervious to water, inks, or other liquids. Surface sizing helps create improved smoothness and surface bonding strength, thereby aiding printability.

Source reduction:
Our first priority in reducing both toxics and waste, by producing and consuming less— reducing the total number and/or size of products and packaging, as well as the use of toxic materials involved in manufacturing, or eliminating any of these things altogether. The reuse of products is crucial to source reduction, since it obviates the need for new ones.

Wastepaper utilization rate:
A rough approximation of the amount of recycled fiber supplying the total content of paper products. Wastepaper utilization rates are calculated by dividing the weight of wastepaper consumed in paper manufacturing by the total weight of paper products produced. Because of fiber loss during processing of wastepaper, the final weight of the recycled pulp obtained will actually be slightly less than the weight of wastepaper originally consumed. Thus the percentage of recycled fiber supplying the final fiber content of the end products will be a little lower than the wastepaper utilization rate.

Appendix 2

Pulping & Papermaking Processes

The various methods for pulping wood are divided into two primary processes—chemical or mechanical. Semi-chemical processes use a combination of chemical and mechanical treatments. In addition to the most commonly used processes summarized here, there are even a few other variations by which limited quantities of wood pulp are manufactured. The characteristics of recycled pulps, which are made by repulping wastepaper, are influenced by the type of virgin pulp originally used to make the paper being repulped.

Pulp is produced by paper mills with integrated pulp and paper manufacturing operations. In addition, there are corporations that do nothing but manufacture market pulp, which they sell to paper mills without their own pulping operations. Some paper mills may also produce enough pulp, in excess of their own capacity to use it, that they may sell some to other companies as market pulp. Typically, market pulp is dried into rough sheets of paper and shipped in bales to customers around the world, where it is made into a wide variety of paper products. Here, market pulp is being shipped from the Portecel terminal in Brazil. (Photo by Jim Young, courtesy of Pulp & Paper *magazine.)*

136 **Recycled Papers: The Essential Guide**

Classification	Process	Description	Yield	Product Use	Total U.S. Production Capacity in 1990

Virgin Wood Pulping Processes

Classification	Process	Description	Yield	Product Use	Total U.S. Production Capacity in 1990
Chemical	Sulfate or Kraft	Over two-thirds of all the wood pulp manufactured in the United States is made by this process. Wood chips are cooked with sodium sulfide and sodium hydroxide under pressure and at raised temperatures, to remove the lignins binding the wood together. The pulp obtained has very good strength properties, and it is used for a great diversity of paper products. Unbleached kraft pulp is dark brown, the color of kraft grocery bags; it is used for bags, wrapping paper, containerboard, and other items. More than one-half of all kraft pulp is bleached for use in high-grade printing, writing, packaging, and other papers. The sulfate pulping process creates sulfur compounds which, when released into the atmosphere, produce the "rotten-egg" smell commonly associated with papermaking.	40–55% of wood by weight	High-grade printing and writing papers, packaging papers, bag and wrapping papers, containerboard, some is added to newsprint.	51 million tons
	Sulfite	Wood chips are cooked in large vessels at increased temperature and pressure with acidic chemicals. Sulfite pulps do not yield paper with as good strength characteristics as sulfate pulps, and the relative importance of this process has declined throughout the 20th century.	45–55% of wood by weight	High-grade printing and writing papers, other publishing papers, some is added to newsprint.	1.5 million tons
Semi-Chemical	Various methods relating to the sulfate, sulfite, or soda processes	Wood chips are subjected to a mild chemical pulping under steam pressure for relatively short periods of time. Subsequently the softened chips are ground up in a manner characteristic of mechanical pulps.	60–80% of wood by weight	Corrugating medium and board, magazine publishing, newsprint, and other products.	5 million tons
Mechanical	Groundwood Thermomechanical (TMP) Chemi-thermomechanical (CTMP) Other variations	Groundwood pulp was the first type of wood pulp developed. It is created by the mechanical grinding of whole logs into fiber, thus giving a high yield of pulp per ton of wood consumed. Thermomechanical pulps are made by first softening wood chips under steam pressure before grinding them. Chemi-thermomechanical pulps use chemicals in addition, usually during the steaming process. Mechanical pulps produce papers with poor sheet strength. Thus they are often combined with chemical pulps in the manufacture of paper products. In general, these papers are prone to fairly rapid deterioration.	90% of wood by weight	Newsprint, groundwood publication papers, tissue and toweling, boxboard, and other papers.	7 million tons

Recycled Paper Pulping Processes

Classification	Process	Description	Yield	Product Use	Total U.S. Production Capacity in 1990
Recycled	Deinking	Wastepaper is repulped and ink, coatings, adhesives, and other contaminants are removed by a variety of processes and equipment. Typically a deinking plant will have a washing and/or flotation stage, in addition to several screening and cleaning stages. In some situations, wastepaper may be repulped and used directly without undergoing an extensive cleaning process. Many different types of paper mills operate deinking plants, including manufacturers of tissue products, newsprint, and paperboard mills. Of the more than fifty deinking plants in North America, only seven are operated by mills making printing and writing paper.	80–90% of the wastepaper consumed	Tissue products, paperboard and containerboard, newsprint, printing and writing papers.	20 million tons (estimate)

Appendix 3

Bibliography & Resources for Further Information

Books

Anderson Fraser Publishing. *Recycled Paper: A Manual for Printers and Designers.* London: Anderson Fraser Publishing, 1990.

Blumberg, Louis, and Robert Gottlieb. *War on Waste: Can America Win Its Battle with Garbage?* Washington, DC: Island Press, 1989.

Casey, James P., ed. *Pulp and Paper Chemistry and Chemical Technology,* 3 vols. New York: John Wiley & Sons, 1980.

Denison, Richard A., and John Ruston, eds. *Recycling & Incineration: Evaluating the Choices.* Washington, DC: Island Press, 1990.

Guthrie, John A. *The Economics of Pulp and Paper.* Pullman, WA: The State College of Washington Press, 1950.

Higham, Robert R. A. *A Handbook of Papermaking.* London: Oxford University Press, 1963.

Hunter, Dard. *Papermaking: The History and Technique of an Ancient Craft* (1943). New York: Dover Publications, 1978.

Katona, George. *The Mass Consumption Society.* New York: McGraw-Hill Book Company, 1964.

McGaw, Judith A. *Most Wonderful Machine: Mechanization and Social Change in Berkshire Paper Making, 1801–1885.* Princeton, NJ: Princeton University Press, 1987.

Miller Freeman Publications. *Pulp & Paper North American Factbook: 1990.* San Francisco: Miller Freeman Publications, 1990.

National Association of Printing Ink Manufacturers. *Printing Ink Handbook.* Harrison, NY: National Association of Printing Ink Manufacturers, 1988.

Packard, Vance. *The Waste Makers.* New York: David McKay Company, 1960.

Saltman, David. *Paper Basics: Forestry Manufacture, Selection, Purchasing, Mathematics and Metrics, Recycling.* New York: Van Nostrand Reinhold Company, 1978.

Scott, William E., in collaboration with Stanley Trosset. *Properties of Paper: An Introduction.* Atlanta, GA: TAPPI Press, 1989.

Smith, David C. *History of Papermaking in the United States: 1691–1969.* New York: Lockwood Publishing Co., 1970.

Sutermeister, Edwin. *The Story of Papermaking.* Boston, MA: S. D. Warren Company, 1954.

Turner, Silvie, and Birgit Skiöld. *Handmade Paper Today.* London: Lund Humphries Publishers, 1983.

Weeks, Lyman Horace. *A History of Paper Manufacturing in the United States, 1690–1916.* New York: Burt Franklin, 1916.

Reports

Attorneys General from California, Florida, Massachusetts, Minnesota, Missouri, New York, Tennessee, Texas, Utah, Washington, Wisconsin. *The Green Report II: Recommendations for Responsible Advertising.* May 1991.

Chandler, William U. *Materials Recycling: The Virtue of Necessity.* Worldwatch Paper 56. Washington, DC: Worldwatch Institute, October 1983.

E. H. Pechan & Associates. *Survey of Deinking Research and Development in the North American Paper and Paperboard Industry.* Springfield, VA: E. H. Pechan & Associates, December 1990.

Franklin Associates, Ltd. *Characterization of Municipal Solid Waste in the United States: 1990 Update.* Washington, DC: U.S. Environmental Protection Agency Office of Solid Waste, June 1990.

Franklin Associates, Ltd. *Paper Recycling: The View to 1995.* New York: American Paper Institute, February 1990.

Franklin Associates, Ltd. *Supply and Recycling Demand for Office Waste Paper, 1990 to 1995.* Washington, DC: National Office Paper Recycling Project, July 1991.

Kroesa, Renate. *The Greenpeace Guide to Paper.* Vancouver, British Columbia: Greenpeace, 1990.

Lester, Stephen, and Brian Lipsett. *Solid Waste Incineration: The Rush to Burn.* Arlington, VA: Citizens Clearinghouse for Hazardous Waste, 1988.

National Solid Wastes Management Association. *Landfill Capacity in the Year 2000.* Washington, DC: National Solid Wastes Management Association, 1989.

Organisation for European Economic Cooperation. *The Pulp and Paper Industry in the USA: A Report by a Mission of European Experts.* Paris: Organisation for European Economic Cooperation, 1951.

Pollock, Cynthia. *Mining Urban Wastes: The Potential for Recycling.* Worldwatch Paper 76. Washington, DC: Worldwatch Institute, April 1987.

U.S. Congress Office of Technology Assessment. *Technologies for Reducing Dioxin in the Manufacture of Bleached Wood Pulp.* OTA-BP-O-54. Washington, DC: U.S. Government Printing Office, May 1989.

U.S. Department of Agriculture Forest Service. *RPA Assessment of the Forest and Rangeland Situation in the United States, 1989.* Forest Resource Report No. 26. Washington DC: U.S. Department of Agriculture, October 1989.

U.S. Department of Agriculture Forest Service. *An Analysis of the Timber Situation in the United States: 1989–2040.* General Technical Report RM-199. Fort Collins, CO: Rocky Mountain Forest and Range Experiment Station, December 1990.

U.S. Environmental Protection Agency. *First Report to Congress: Resource Recovery and Source Reduction.* Washington, DC: U.S. Environmental Protection Agency, 3rd ed., 1974.

Wiseman, A. Clark. *U.S. Wastepaper Recycling Policies: Issues and Effects.* Washington, DC: Resources for the Future, Energy and Natural Resources Division, August 1990.

Conference Proceedings & Symposia

Arthur, Jett C., Jr., ed. *Cellulose and Fiber Science Developments: A World View.* Washington, DC: American Chemical Society, 1977.

CERMA. *First National Symposium on Recycled Paper.* Atlanta, GA: Center for Earth Resource Management Applications, June 1990. (No proceedings published.)

CERMA. *Second National Symposium on Recycled Paper.* St. Louis, MO: Center for Earth Resource Management Applications, April 1991. (No proceedings published.)

Focus '95+. *Proceedings of the Landmark Paper Recycling Symposium.* Atlanta, GA: TAPPI Press, 1991.

New England TAPPI/Connecticut Valley Pima. Technical Seminar: *Meeting the Recycle Challenge.* Holyoke, MA: New England TAPPI/Connecticut Valley Pima, March 1991. (No proceedings published.)

Pulp & Paper. *Proceedings of Wastepaper I: Demand in the 90's.* Chicago, IL: Miller Freeman Publications, May 1990.

Pulp & Paper. *Proceedings of Wastepaper II: Markets & Technologies.* Chicago, IL: Miller Freeman Publications, May 1991.

Exhibition Catalogs

Library of Congress. *Papermaking: Art and Craft.* Washington, DC: Library of Congress, 1968.

New York Public Library. *On Paper: The History of an Art.* Exhibition curated by Robert Rainwater. New York: New York Public Library, 1990.

New York Public Library. *On Paper.* Letterpress edition. New York: New York Public Library, 1990.

Smithsonian Institution, National Museum of American History. *300 Years of American Papermaking.* Exhibition curated by Helena E. Wright. Washington, DC: Smithsonian Institution, 1991.

Reference Books

Buy Recycled! Your Practical Guide to the Environmentally Responsible Office. Chicago, IL: Services Marketing Group, 1990.

CERMA'S Recycled Paper Handbook, 1st ed. Springfield, VA: The Center for Earth Resource Management Applications, 1991.

The Competitive Grade Finder, 24th ed. Exton, PA: Grade Finders, Inc., 1990.

The Dictionary of Paper, 4th ed. New York: American Paper Institute, 1980.

Paper Buyers Index System. Phoenixville, PA: Printing Complaints, Inc., 1990.

Paper Complaint Handbook System. Valley Forge, PA: Printing Complaints Inc., 1986.

Paper, Paperboard, and Wood Pulp Capacity. New York: American Paper Institute, 1990.

Papermatcher™ A Directory of Paper Recycling Resources. New York: American Paper Institute, 1990.

Recycled Printing & Writing Papers: Products & Manufacturers. Springfield, VA: Center for Earth Resource Management Applications, June 1991.

Statistics of Paper, Paperboard & Wood Pulp: 1990. New York: American Paper Institute, 1990.

Walden's Handbook for Paper Salespeople & Buyers of Printing Paper, 2nd ed. Oradell, NJ: Walden-Mott Corporation, 1981.

Periodicals

Alkaline Paper Advocate. Abbey Publications, 320 East Center Street, Provo, UT 84606. TEL 801-373-1598.

BioCycle: Journal of Waste Recycling. The JG Press, Box 351, 18 South Seventh Street, Emmaus, PA 18049. TEL 215-967-4135.

Fibre Market News. 4012 Bridge Avenue, Cleveland, OH 44113. TEL 800-456-0707.

Garbage: The Practical Journal for the Environment. Old House Journal Corp., 2 Main Street, Gloucester, MA 01930. TEL 508-283-3200.

The Official Recycled Products Guide. Recycled Products Guide, P.O. Box 577, Ogdensburg, NY 13669. TEL 800-267-0707.

PaperAge. Global Publications, 400 Old Hook Road, Westwood, NJ 07675. TEL 201-666-2262.

Paper Recycler. Miller Freeman Publications, 600 Harrison Street, San Francisco, CA 94107. TEL 415-905-2371.

The Paper Stock Report. 13727 Holland Road, Cleveland, OH 44142. TEL 216-362-7979.

PIMA Magazine. Paper Industry Management Association, 2400 East Oakton Street, Arlington Heights, IL 60005. TEL 708-956-0250.

Pulp & Paper. Miller Freeman Publications. 600 Harrison Street, San Francisco, CA 94107. TEL 415-267-7645.

Pulp & Paper Project Report. Miller Freeman Publications, 600 Harrison Street, San Francisco, CA 94107. TEL 415-905-2371.

Recycled Grade Finder: A Quarterly Service. Jaakko Pöyry Consulting, 560 White Plains Road, Tarrytown, NY 10591-5136. TEL 914-332-4000.

Recycled Paper News. Center for Earth Resource Management Applications, 5528 Hempstead Way, Springfield, VA 22151. TEL 703-642-1120 ext. 116.

Recycling Times. Recycling Times, 5615 West Cermak Road, Cicero, IL 60650. TEL 202-861-0708.

Resource Recycling. Resource Recycling, Inc., P.O. Box 10540, Portland, OR 97209. TEL 800-227-1424 or 503-227-1319.

TAPPI Journal. Technical Association of the Pulp and Paper Industry, 15 Technology Parkway South, Norcross, GA 30092. TEL 404-446-1400.

Waste Age. National Solid Wastes Management Association, 1730 Rhode Island Avenue NW, Suite 1000, Washington, DC 20036. TEL 202-659-4613.

Organizations

American Institute of Graphic Arts
1059 Third Avenue, New York, NY 10021
TEL 212-752-0813

American National Standards Institute (ANSI)
11 West 42nd Street, 13th Floor, New York, NY 10036
TEL 212-642-4900

American Paper Institute (API)
260 Madison Avenue, New York, NY 10016
TEL 212-340-0600

American Society for Testing and Materials (ASTM)
1916 Race Street, Philadelphia, PA 19103-1187
TEL 215-299-5400

Californians Against Waste Foundation, Buy Recycled Campaign
909 12th Street, Suite 201, Sacramento, CA 95814
TEL 916-443-8317

Citizens Clearinghouse for Hazardous Waste (CCHW)
P.O. Box 6806, Falls Church, VA 22040
TEL 703-237-2249

Conservation Law Foundation (CLF)
3 Joy Street, Boston, MA 02108-1497
TEL 617-742-2540

Conservatree Paper Company
10 Lombard Street, Suite 250, San Francisco, CA 94111
TEL 415-433-1000

Environmental Choice / EcoLogo
Environment Canada
107 Sparks Street, Second Floor, Ottawa, Ontario K1A OH3
TEL 613-952-9440

Environmental Defense Fund (EDF)
257 Park Avenue South, New York, NY 10010
TEL 212-505-2100

Environmental Protection Agency (EPA)
Procurement Guidelines Office
Municipal Solid Waste Program, Mailcode OS-301
401 M Street SW, Washington, DC 20460
TEL 202-382-2780

Environmental Protection Agency (EPA)
Procurement Guidelines Hotline / Recycled Products Information
c/o CERMA, 5528 Hempstead Way, Springfield, VA 22151
TEL 703-941-4452

Federal Trade Commission (FTC)
Division of Advertising Practices
6th and Pennsylvania Avenue NW, Washington, DC 20580
TEL 202-326-3090

Green Cross Certification Company
1611 Telegraph Avenue, Suite 1111, Oakland, CA 94612-2113
TEL 415-832-1415

Green Seal
P.O. Box 1694, Palo Alto, CA 94302-1694
TEL 415-327-2200

Greenpeace
1436 U Street NW, Washington, DC 20009
TEL 202-462-1177

Institute for Local Self-Reliance
2425 18th Street NW, Washington, DC 20009
TEL 202-232-4108

Institute of Paper Science and Technology
575 14th Street NW, Atlanta, GA 30318
TEL 404-853-9500

Institute of Scrap Recycling Industries (ISRI)
Paper Stock Institute
1627 K Street NW, Suite 700, Washington, DC 20006
TEL 202-466-4050

National Office Paper Recycling Project
The United States Conference of Mayors
1620 Eye Street NW, Washington, DC 20006
TEL 202-293-7330

National Paper Trade Association
111 Great Neck Road, Great Neck, NY 11021
TEL 516-829-3070

National Recycling Coalition (NRC)
1101 30th Street NW, Washington, DC 20007
TEL 202-625-6406

Natural Resources Defense Council (NRDC)
40 West 20th Street, New York, NY 10011
TEL 212-727-2700

Northeast Recycling Council (NERC)
139 Main Street, Suite 401, Brattleboro, VT 05301
TEL 802-254-3636

Technical Association of the Pulp and Paper Industry (TAPPI)
P.O. Box 105113, Atlanta, GA 30348
TEL 800-332-8686

The Wilderness Society
900 Seventeenth Street NW, Washington, DC 20006-2596
TEL 202-833-2300

Worldwatch Institute
1776 Massachusetts Avenue NW, Washington, DC 20036
TEL 202-452-1999

Appendix 4

Recycled Papers Available

This listing of recycled papers was compiled by means of extensive telephone interviews with representatives of companies that make or sell printing and writing papers in the United States and Canada. We attempted to include every manufacturer that markets recycled papers, and we discussed our survey with fifty-five different corporations or corporate divisions. In order to have their papers included in this list, we asked the mill representatives to answer a standardized set of questions about each recycled paper grade sold. Very few companies declined to participate in our survey. We appreciate the cooperation of all who provided us with information.

We report two recycled content percentages for each paper; both are based on total fiber content rather than on total paper weight. All percentages listed represent the minimum recycled content routinely used to make that particular paper. The *Total Recycled Fiber Content* figure includes recycled fiber from all sources— ranging from internal mill wastes to heavily printed paper wastes purchased from independent sources, as well as everything in between. A few mills exclude internally generated mill wastes in their total recycled content figures, reporting only postmill purchased fiber in this percentage. Unfortunately, we are unable to differentiate between the mills using these two different reporting methods. We do note papers that have been certified by either the Canadian EcoLogo program or the New York State recycled product certification program.

Based on our conclusions that definitions of postconsumer content are not being consistently applied from mill to mill, we chose not to report existing postconsumer content claims. Instead, we asked for the specific percentage of fiber content supplied by wastepaper that has actually been processed through a deinking plant— the *Deinked Recycled Fiber Content*. While this fiber is already included in the total recycled fiber content, this figure gives additional information about the types of wastes used in manufacturing. In addition, a few mills use printed paper wastes without deinking them. In these cases, we put the code (p) under the heading, *Non-deinked Printed Wastes*. This code does not specify the percentage of non-deinked printed wastepaper used. However, we only use this notation when these wastes supply at least 20% of the fiber furnish.

When deinked recycled fiber is used, a lower percentage is sometimes used in the manufacture of white papers than in the darker colored papers of the same grade; in these cases we listed the content percentage as a range, with the lower percentage applying to the whites and the higher percentage applying to the colors. The fact that this percentage may vary was volunteered by some of the mills we spoke with. Similar variations may exist for others of the papers listed here, but we may not have been given that information in all cases.

The *Cotton Fiber Content* is reported as a separate percentage. This figure is not included in the total recycled fiber content.

In the course of our survey, we often had to speak with several representatives of the same company or mill to have our questions answered. Quite frankly, we often got different answers to the same questions depending on with whom we were speaking. When answers differed, we tried to follow up with additional questions in order to come to a reasonable conclusion about what seemed accurate. As much as possible, we spoke directly with managing superintendents or technical directors who oversee the daily manufacturing operations, and are thus in a position to answer the very specific questions we asked. A few corporations, however, would only allow us to speak with their sales staff.

Some manufacturers market both virgin and recycled papers under the same product name, making it difficult to distinguish which of those papers are recycled. For example, a particular grade may have some colors which are made with recycled fibers and others which are not, and this may not be readily evident in the promotional materials. Some papers may have been omitted because information was inconsistent or complete information was not available to us.

What is reported here is based solely on the information given to us by management employees at each company. Because we are not a certifying agency, we cannot verify its accuracy. In addition, manufacturing formulas for existing papers often change, new grades are introduced, and the use of recycled fiber, in particular, is in a great deal of flux. Thus we urge readers to ask their own questions and check all information with the paper mills and their own suppliers. We have made every effort to conduct a detailed and meaningful survey, and to report the data we were given accurately and in a consistent and fair manner. Given that there are widely differing definitions, standards, and reporting procedures at the different mills, this was not an easy task. We apologize for any errors that may be present.

Key

Certifications:	EL	=	certified under the Canadian EcoLogo Program
	NYS	=	certified by the New York State Department of Environmental Conservation

Paper Classifications: Paper classifications vary widely. We have summarized the classifications as follows:

- **offset:** any papers used in offset or other printing applications
- **writing:** bond, writing, or tablet papers
- **copier:** photocopy, laser printer, or other duplicating papers
- **bristol:** index or bristol papers

We have also listed some specialty papers according to their intended use.

Fiber Content Percentages: All percentages are based on total fiber content rather than on total paper weight. For each paper, the total recycled fiber content percentage includes the deinked fiber content, but not the cotton fiber content. If the deinked recycled fiber content varies with the color of the paper made, we list this percentage as a range.

(p) identifies papers in which printed wastepaper has been used in the furnish directly without deinking, and in which it supplies at least 20% of the total fiber content.

Certifications	Paper Company & Grade	Paper Classification	Color Palette	Finishes	Alkaline with Minimum pH 7.5	Total Recycled Fiber Content	Deinked Recycled Fiber Content	Non-deinked Printed Wastes	Cotton Fiber Content
	Coated Free-sheet								
	Ahlstrom Paper Company								
	MasterArt R/C	offset	whites only	dull, gloss		50%	20%		0%
	Champion International								
	Kromekote 2000 Recycled	offset	whites only	cast coated		50%	10%		0%
	Conservatree								
	Conservatree C1S/C2S	offset	whites only	gloss		60%	10%		0%
	Conservatree Offset Enamel	offset	whites only	gloss		60%	10%		0%
	Domtar Specialty Fine Papers								
	Cornwall Coated Cover	offset	whites only	smooth		50%	0%		0%
	P.H. Glatfelter Company								
	EPA Enamel/Matte	offset	whites only	gloss, matte	•	55%	50%		0%
	Honshu Paper Company								
	Lotus Gloss/Matte	offset	whites only	gloss, matte		80%	80%		0%
	Island Paper Mills								
	ReSolve Coated	offset	whites only	gloss	•	50%	0%		0%
	Mohawk Paper Mills								
EL/NYS	Mohawk 50/10	offset	whites only	gloss, matte	•	60%	0%		0%
	Noranda Forest Recycled Papers								
EL	Regeneration Cover C1S	offset	whites only			75%	50%		0%
	Potlatch Corporation								
	Quintessence Remarque	offset	whites only	gloss	•	50%	0%		0%
	Simpson Paper Company								
	EverGreen Gloss/Matte	offset	whites only	gloss, matte	•	60%	0%		0%
	Satin-Kote Recycled	offset	whites only	matte	•	60%	0%		0%
	Stora Papyrus Newton Falls								
	Champlain Gloss	offset	whites only	gloss	•	50%	0%		0%
	St. Lawrence Gloss/Matte	offset	whites only	gloss, matte	•	50%	0%		0%

Certifications	Paper Company & Grade	Paper Classification	Color Palette	Finishes	Alkaline with Minimum pH 7.5	Total Recycled Fiber Content	Deinked Recycled Fiber Content	Non-deinked Printed Wastes	Cotton Fiber Content
	S.D. Warren								
	Recovery Gloss/Matte	offset	whites only	gloss, matte	•	50%	25%		0%
	Westvaco								
	American Eagle Web	offset	whites only	dull, gloss, matte	•	50%	0%		0%

Uncoated Free-sheet

Certifications	Paper Company & Grade	Paper Classification	Color Palette	Finishes	Alkaline with Minimum pH 7.5	Total Recycled Fiber Content	Deinked Recycled Fiber Content	Non-deinked Printed Wastes	Cotton Fiber Content
	Badger Paper Mills								
	Envirographic	copier	whites & colors	smooth		50%	10%		0%
	Beckett Paper Company								
	Beckett Ridge	offset	whites & colors	deeply embossed		50%	10%		0%
	Cambric Writing	offset, writing	whites & colors	linen		50%	10%		0%
	Concept	offset, writing	whites & colors	linen, wove	•	50%	10%		0%
	Enhance!	offset, writing	whites & colors	deeply embossed satin		50%	10%		0%
	RSVP	offset	whites & colors	felt		50%	25%		0%
	Beveridge Paper Company								
	Rainbow Bristol	bristol	whites & colors	smooth	•	100%	0%		0%
	Boise Cascade								
	Aspen Offset	offset	whites & colors	smooth, vellum	•	50%	30%		0%
	Aspen Xerographic	copier	whites & colors	smooth	•	50%	30%		0%
	RC Form Bond	writing	whites & colors	bond	•	50%	30%		0%
	RC Summit Ledger	writing	whites & colors	bond	•	50%	30%		0%
	RC Wove Envelope	offset, copier	whites & colors	smooth	•	50%	30%		0%
	Recycled Tablet	writing	whites & colors	tablet	•	50%	30%		0%
	Recycled Tag	bristol	whites only	bond	•	50%	30%		0%
	3800 Recycled Bond	writing, copier	whites only	bond	•	50%	30%		0%
	CPM								
EL	NVF (Non Virgin Fibers)	offset (envelope)	whites & colors	wove		100%	0%		0%
	Champion International								
	Benefit Text & Cover/Writing	offset, writing	whites & colors	felt, linen, vellum		50%	0–20%		0%
	Benefit Natural	offset, writing	colors only	vellum		50%	0%	(p)	0%
	Conservatree								
	Conservatree 100	offset	whites & colors	vellum		100%	100%		0%
	Conservatree Bond	writing, copier	whites only	smooth		60%	10%		0%
	Conservatree Computer Paper	copier	white & greenbar	smooth		65%	65%		0%
	Conservatree Four Seasons	offset	colors only	felt		75%	15%		0%

Certifications	Paper Company & Grade	Paper Classification	Color Palette	Finishes	Alkaline with Minimum pH 7.5	Total Recycled Fiber Content	Deinked Recycled Fiber Content	Non-deinked Printed Wastes	Cotton Fiber Content
	Conservatree (continued)								
	Conservatree Premium Opaque	offset	whites only	satin, vellum		100%	100%		0%
	Conservatree Premium Rag Bond	offset, writing	whites & off-whites	cockle, laid, linen		75%	20%		25%
	Conservatree Premium Xerographic	copier	whites only	smooth		60%	60%		0%
	Crane & Company								
	Crane's Crest Recycled	writing	whites & off-whites	smooth		*50%	0%		100%
	*Includes some postconsumer cotton wastes from discarded sheets, tablecloths, and other cotton products.								
	Cross Pointe								
	Bellbrook Laid	offset	whites & colors	laid	•	70%	50%		0%
	Cross Pointe Book	offset	whites & off-whites	smooth, vellum	•	70%	50%		0%
	Cross Pointe Brights	offset	colors only	vellum	•	50%	0%		0%
	Cross Pointe Recycled Bond	offset, writing, copier	whites & off-whites	antique, cockle, smooth	•	50%	50%		25%
	Flambeau Vellum Bristol & Index	bristol	whites & colors	vellum	•	50%	10%		0%
EL	Genesis	offset, writing	colors only	smooth, vellum	•	100%	100%		0%
	Halopaque	offset	whites & off-whites	satin, vellum	•	70%	50%		0%
	Heritage Book	offset	whites & off-whites	antique, smooth, vellum	•	70%	50%		0%
	Medallion	offset	whites & colors	felt	•	70%	40–50%		0%
	Miami Book	offset	whites & off-whites	antique, satin, smooth, vellum	•	70%	50%		0%
	Normandie	offset, writing	whites & colors	linen	•	70%	40–50%		0%
	Passport	offset, writing	whites & colors	felt, smooth	•	70%	40–50%		0%
	Sycamore Colors	offset	colors only	smooth, vellum	•	70%	50%		0%
	Torchglow Opaque	offset	whites & colors	smooth	•	50%	15%		0%
	Troy Book	offset	whites & off-whites	antique, satin, smooth, vellum	•	70%	50%		0%
	Domtar Specialty Fine Papers								
EL	Byronic Recycled	offset, writing	whites & colors	brocade		70%	0%		0%
EL	Concerto Recycled Text & Cover	offset	whites & colors	wove		70%	0%		0%
EL	Concerto Recycled Writing	writing	whites & colors	wove		70%	0%		25%
EL	Mayfair Recycled Cover	offset	whites & colors	antique		60%	0%		0%
EL	Natural Cover 100	offset	one color	smooth		100%	0%	(p)	0%
EL	Plainfield Recycled Offset	offset	whites & colors	smooth		60%	0%		0%
EL	Sandpiper	offset, writing	whites & colors	wove		100%	0%	(p)	0%
	Eastern Fine Paper								
	Atlantic Bond	writing	whites only	bond	•	50%	0%		0%
	Atlantic Opaque	offset	whites & colors	deep etch, regular, vellum	•	50%	0–10%		0%
	Bar Harbor Felt	offset	whites & colors	felt	•	50%	0%		0%
	Certificate Bond II/Laser II	writing, copier	whites only	cockle, smooth	•	50%	0%		25%
	Certificate Ledger	writing	whites & colors	smooth	•	50%	0%		25%
	Certificate Linen	offset, writing	whites & colors	linen	•	50%	0%		25%
	Certificate Royale	offset, writing	whites & colors	laid	•	50%	0%		25%

Certifications	Paper Company & Grade	Paper Classification	Color Palette	Finishes	Alkaline with Minimum pH 7.5	Total Recycled Fiber Content	Deinked Recycled Fiber Content	Non-deinked Printed Wastes	Cotton Fiber Content
	Eastern Fine Paper (continued)								
	Eastern Opaque	offset	whites only	laid, linen, regular, vellum	•	50%	10%		0%
	Imperial Character Bond	writing	whites only	bond	•	50%	0%		50%
	Monhegan Canvas	offset	whites & colors	canvas	•	50%	0%		0%
	Premium Bristol Card Stock	bristol	whites only	smooth	•	50%	0%		0%
	Esleek Manufacturing								
	Reissue Bond	writing	whites only	wove		75%	15%		25%
	Fletcher Paper Company								
	Geopake	writing	whites only	smooth, vellum		50%	10%		0%
	Recycled Financial	writing	whites only	smooth		50%	10%		0%
	Fox River Paper Company								
EL	Fox River Bond	offset, writing, copier	whites only	light cockle, smooth	•	75%	20%		25%
	Circa '83	offset, writing	whites & colors	laid		55%	22%		0%
	Circa Select	offset, writing	whites & colors	antique, writing		55%	22%		0%
	Confetti	offset	colors only	smooth		55%	27.5%		0%
	Early American	offset	whites & colors	felt		55%	27.5%		0%
	French Paper Company								
	French Linen	offset	whites & colors	linen		50%	0%		0%
	French Rayon	offset	whites & colors	vellum	•	75%	15%		0%
	Speckletone	offset	whites & colors	antique, cordtone, versailles	•	75%	15%		0%
	Georgia-Pacific								
	Ardor Recycle Xero/Bond	copier	whites & colors	bond		50%	15%		0%
	Gilbert Paper								
	Esse	offset, writing	whites & colors	smooth, texture		50%	15%		0%
	Gilbert Recycled	offset, writing	whites & off-whites	cockle, wove		50%	15%		25%
	P.H. Glatfelter Company								
	Norse Cote	offset	whites only	satin	•	55%	50%		0%
	Odin Offset	offset	whites only	eggshell, regular	•	55%	50%		0%
	Recycle/100 Offset	offset	colors only	regular	•	100%	0%		0%
	Restore Cote	offset	whites only	satin	•	55%	50%		0%
	Hopper Paper Company								
EL	Proterra	offset, writing	whites & colors	felt, laid, linen, vellum		58%	58%		0%
	Howard Paper/Division of Fox River								
	Capitol Bond	writing, copier	whites & off-whites	light cockle, laser		50%	0%		25%
	Howard Linen	offset, writing	whites & colors	linen		50%	0%		0%

Appendix 4: Recycled Papers Available

Certifications	Paper Company & Grade	Paper Classification	Color Palette	Finishes	Alkaline with Minimum pH 7.5	Total Recycled Fiber Content	Deinked Recycled Fiber Content	Non-deinked Printed Wastes	Cotton Fiber Content
	IP/Hammermill Division								
	Savings DP	writing, copier	whites & colors			50%	10%		0%
	Savings Offset	offset	whites only	smooth, vellum	•	50%	0%		0%
	Savings Opaque	offset	whites only	smooth		50%	10%		0%
	Island Paper Mills								
EL	ReSolve Bond/Copy/Register	writing, copier	whites & colors	regular	•	50%	0%		0%
EL	ReSolve Offset	offset	whites only	regular	•	50%	0%		0%
	James River								
	Curtis Brightwater Marble	offset	colors only	smooth	•	60%	20%		0%
	Curtis Tuscan Antique	offset	whites & colors	antique, pab		60%	0%		0%
	Curtis Tuscan Terra	offset	whites & colors			60%	0%		0%
	Discover Jersey Cover	offset	whites & colors	highlight, leatherette, plate		50%	0%		0%
	Graphika 100	offset, copier	whites & colors	vellum	•	100%	0%	(p)	0%
	Retreeve	offset	whites & colors	felt, vellum		90%	0%		0%
	Retreeve Writing	writing	whites & colors	laid, wove		60%	0–20%		0%
	Riegel GC Cover	offset	colors only	vellum		50%	0%		0%
	Riegel PCW Cover	offset	whites & colors	vellum		70%	0%	(p)	0%
	Little Rapids Paper Division/Potsdam Paper Mills								
	Potsdam GPO Grades	offset	whites only	smooth		50%	0%		0%
	Potsdam Recycled	offset, writing	whites & colors	smooth, vellum, wove		50%	0%		0%
	Lyons Falls								
	Paper Again PC	offset, writing	whites & colors	antique, smooth, vellum	•	50%	0%		0%
	Mead Corporation								
	Harmony Recycled Xerographic	copier	whites only			50%	0%		0%
	Mohawk Paper Mills								
EL/NYS	Mohawk P/C Whites	offset	whites & off-whites	felt, linen, vellum	•	60%	0%		0%
EL/NYS	Mohawk P/C Colors	offset	colors only	felt, linen, vellum	•	100%	0%	(p)	0%
EL/NYS	Mohawk P/C 100	offset	gray	vellum	•	100%	0%	(p)	0%
	Monadnock Paper Mills								
	Revue	offset	whites & off-whites	smooth	•	60%	15%		0%
	Neenah Paper								
	Environment	offset	whites & colors	laid, linen, parchment, wove		100%	15%		0%
	Environment 25 Writing	writing	whites & colors	laid, linen, parchment, wove		75%	15%		25%

Certifications	Paper Company & Grade	Paper Classification	Color Palette	Finishes	Alkaline with Minimum pH 7.5	Total Recycled Fiber Content	Deinked Recycled Fiber Content	Non-deinked Printed Wastes	Cotton Fiber Content
	Noranda Forest Recycled Papers								
EL	Oxford Opaque	offset	whites only	english, regular, vellum	•	50%	50%		0%
EL	Phoenix Opaque	offset	whites & off-whites	english, regular, vellum	•	50%	50%		0%
EL	Renaissance Reply Card	offset	whites only			75%	50%		0%
	Patriot Paper								
	Patriot Copier Bond	copier	whites only	bond	•	100%	40%		0%
	Patriot Cover	offset	whites only	vellum	•	100%	40%		0%
	Patriot Index	bristol	whites only	smooth	•	100%	40%		0%
	Patriot Offset	offset, copier	whites only	antique, regular, vellum	•	100%	40%		0%
	Patriot Reply Card	bristol	whites only	antique	•	100%	40%		0%
	Riverside								
	Ecology Bond/Offset	offset, writing, copier	whites & colors	smooth	•	100%	10%		0%
	Sav-A-Source								
	Sav-A-Source Writing	writing	whites & off-whites	smooth		50%	0%		25%
	Simpson Paper Company								
	Coronado SST Recycled	offset, bristol	whites only	modified antique, stipple		50%	0%		0%
	Ecopaque	offset	whites & colors	vellum		50%	33%		0%
	EverGreen	offset, writing	whites & colors	vellum		50%	10%		0%
	Quest	offset	whites & colors	machine		100%	0%	(p)	0%
	Simpson Emblem Xerographic	copier	whites only	bond		50%	33%		0%
	Simpson Recycled Offset	offset	whites only	wove		50%	33%		0%
	Simpson Recycled Xerographic	copier	whites only	bond		50%	33%		0%
	Strathmore Paper Company								
	Americana	offset	whites & colors	felt		50%	0%		0%
	Renewal Text & Cover	offset	whites & colors	wove	•	50%	25%		0%
	Renewal Writing	writing	whites & colors	wove	•	50%	25%		25%
	Strathmore Bond	writing, copier, bristol	whites & off-whites	cockle, laid, smooth, wove		25%	0%		25%
	Ward Paper Company								
	Cimarron	offset	colors only	vellum		50%	25%		0%
	Forward Vellum Opaque	offset	colors only	vellum		50%	0%		0%
	Lake Shore 25	writing, copier	whites & off-whites	wove		50%	20%		25%
	Wausau Paper Company								
	Royal	offset, writing	whites & colors	bond, felt, linen, vellum		50%	10%		0%
	Westvaco								
	American Eagle Bulking Offset	offset	whites only	smooth	•	50%	0%		0%
	American Eagle Envelope Wove	offset	whites only	smooth	•	50%	0%		0%

Certifications	Paper Company & Grade	Paper Classification	Color Palette	Finishes	Alkaline with Minimum pH 7.5	Total Recycled Fiber Content	Deinked Recycled Fiber Content	Non-deinked Printed Wastes	Cotton Fiber Content
	Weyerhaeuser Paper								
	Recycled Laser Copy	copier	whites only		•	50%	10%		0%
	Recycled Lynx Opaque	offset	whites only	smooth, vellum	•	50%	10%		0%
	George A. Whiting Paper Company								
	Brockway	offset	whites & colors	felt	•	100%	10%		0%
	Cadence	offset, writing	whites & colors	linen	•	100%	10%		0%
	Coat of Arms	offset	whites & colors	antique, embossed	•	100%	10%		0%
	Crestline	offset	whites & colors	vellum, embossed	•	100%	10%		0%
	Polar White Bristol	bristol	whites only	smooth	•	100%	10%		0%
	Ultima	offset, writing	whites & colors	laid	•	100%	10%		0%

Coated Groundwood

Certifications	Paper Company & Grade	Paper Classification	Color Palette	Finishes	Alkaline with Minimum pH 7.5	Total Recycled Fiber Content	Deinked Recycled Fiber Content	Non-deinked Printed Wastes	Cotton Fiber Content
	Champion International								
	Sunweb Recycled	offset	whites only	gloss		50%	10%		0%
	Textweb Recycled	offset	whites only	gloss		50%	10%		0%
	Conservatree								
	ESP Gloss	offset	whites only	gloss		75%	75%		0%
	Consolidated Papers								
	ConCycle #4/#5	offset	whites only	gloss		50%	0%		0%
	Georgia-Pacific								
	Re-Comm Matte	offset	whites only	matte		50%	30%		0%
	Niagara of Wisconsin Paper Company								
	Pentair Recycle Gloss/Suede	offset	whites only	gloss, suede		54%	34%		0%
	Simpson Paper Company								
	Nature Web	offset	whites only	gloss, suede		50%	10%		0%

Uncoated Groundwood

Certifications	Paper Company & Grade	Paper Classification	Color Palette	Finishes	Alkaline with Minimum pH 7.5	Total Recycled Fiber Content	Deinked Recycled Fiber Content	Non-deinked Printed Wastes	Cotton Fiber Content
	FSC Paper Company								
	FSC E-Grade	writing	off-whites only	machine		100%	100%		0%
	Tablet	writing	colors only	machine		100%	100%		0%
	Georgia-Pacific								
	Re-Comm Offset	offset	off-whites & colors	smooth, vellum		50%	30%		0%
	Re-Run Bond	writing	off-whites & colors	bond		50%	30%		0%

Certifications	Paper Company & Grade	Paper Classification	Color Palette	Finishes	Alkaline with Minimum pH 7.5	Total Recycled Fiber Content	Deinked Recycled Fiber Content	Non-deinked Printed Wastes	Cotton Fiber Content
	P.H. Glatfelter Company								
	Recolocote	offset	off-whites only	satin		55%	50%	0%	
	Hennepin Paper Company								
	Fibertone Mimeo	copier	off-whites & colors	vellum		50%	0%	0%	
	Manistique Papers								
	Groundwood Offset/Insert	offset	off-whites & colors	smooth	•	100%	100%	0%	
	Patriot Paper								
	NP-100	offset, writing, bristol	off-whites only	antique, bond, vellum	•	100%	50%	0%	
	Steinbeis								
	Recyconomic	copier	off-whites & colors			100%	50%	0%	
	Recyconomic	offset	off-whites & colors			100%	90%	0%	

Specialty Papers

Certifications	Paper Company & Grade	Paper Classification	Color Palette	Finishes	Alkaline with Minimum pH 7.5	Total Recycled Fiber Content	Deinked Recycled Fiber Content	Non-deinked Printed Wastes	Cotton Fiber Content
	Appleton Papers								
	Optima Recover	fax	whites only	heat sensitive coating	•	50%	50%	0%	
	Recover	carbonless	whites & colors	specialty coating	•	50%	50%	0%	
	Badger Paper Mills								
	SHARPrint	computer	white & greenbar	bond		50%	10%	0%	
	Boise Cascade								
	Recycled Safety	bank check	patterned	smooth	•	50%	30%	0%	
	Hennepin Paper Company								
	NRA Target (groundwood)	target paper	off-white	vellum		50%	0%	0%	
	Album Stock (groundwood)	photo album	black	antique		50%	0%	0%	
	Mead Corporation								
	Sequel	carbonless	whites only	specialty coating		50%	10%	0%	
	3M								
	Scotchmark Recycled	carbonless	whites & colors	specialty coating		59%	31%	0%	

Appendix 5

Designer Impact Analysis Form

See *The Paper Specifier's Impact* in chapter 6 for a discussion of the significance of the paper specifier's role. To use this form follow the instructions below.

Part 1

Fill in the descriptive information recording each paper specified— name, classification, basis weight, color, and sheet size used for printing. Fill in the number of cartons ordered by the printer, and the sheets per carton and M weights (information found in the swatch book).

Multiply the number of cartons by the sheets per carton and by the M weights, and then divide by 1000, to calculate the total pounds of each paper specified.

Enter the percentage recycled fiber content as a decimal figure (for example, 50% = 0.50), for each paper. Multiply the total pounds of each paper specified by the percentage recycled fiber content, to obtain the pounds of paper specified equivalent to 100% recycled paper. Enter in boxes A, B, C, and D.

Add (A+B+C+D) to obtain the total pounds of paper specified equivalent to 100% recycled paper, and enter in box E.

Part 2

Divide the pounds specified by 2000, to obtain the total tons of 100% recycled paper specified, and enter in box F.

Part 3

Multiply the total tons of 100% recycled paper by each of the conversion factors for trees saved, landfill space saved, and air pollutants avoided.

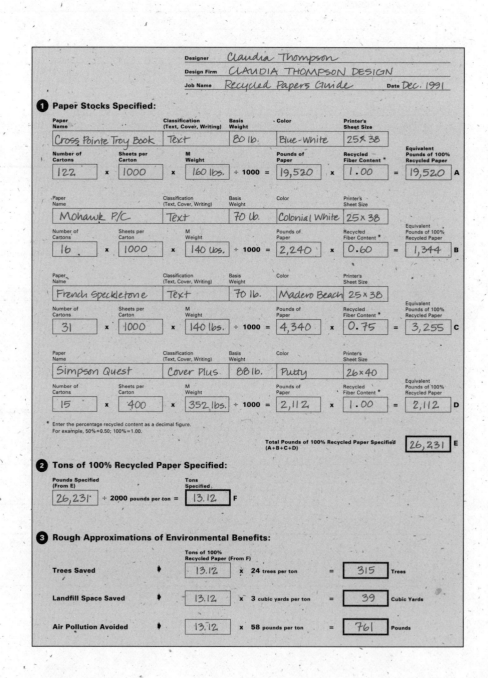

156 Recycled Papers: The Essential Guide

Designer _____

Design Firm _____

Job Name _____ Date _____

1. Paper Stocks Specified:

Paper Name | **Classification** (Text, Cover, Writing) | **Basis Weight** | **Color** | **Printer's Sheet Size**

Number of Cartons [ ] x **Sheets per Carton** [ ] x **M Weight** [ ] ÷ 1000 = **Pounds of Paper** [ ] x **Recycled Fiber Content*** [.] = **Equivalent Pounds of 100% Recycled Paper** [ ] **A**

Paper Name | **Classification** (Text, Cover, Writing) | **Basis Weight** | **Color** | **Printer's Sheet Size**

Number of Cartons [ ] x **Sheets per Carton** [ ] x **M Weight** [ ] ÷ 1000 = **Pounds of Paper** [ ] x **Recycled Fiber Content*** [.] = **Equivalent Pounds of 100% Recycled Paper** [ ] **B**

Paper Name | **Classification** (Text, Cover, Writing) | **Basis Weight** | **Color** | **Printer's Sheet Size**

Number of Cartons [ ] x **Sheets per Carton** [ ] x **M Weight** [ ] ÷ 1000 = **Pounds of Paper** [ ] x **Recycled Fiber Content*** [.] = **Equivalent Pounds of 100% Recycled Paper** [ ] **C**

Paper Name | **Classification** (Text, Cover, Writing) | **Basis Weight** | **Color** | **Printer's Sheet Size**

Number of Cartons [ ] x **Sheets per Carton** [ ] x **M Weight** [ ] ÷ 1000 = **Pounds of Paper** [ ] x **Recycled Fiber Content*** [.] = **Equivalent Pounds of 100% Recycled Paper** [ ] **D**

* Enter the percentage recycled content as a decimal figure. For example, 50%=0.50; 100%=1.00.

Total Pounds of 100% Recycled Paper Specified (A+B+C+D) [ ] **E**

2. Tons of 100% Recycled Paper Specified:

Pounds Specified (From E) [ ] ÷ 2000 pounds per ton = **Tons Specified** [ ] **F**

3. Rough Approximations of Environmental Benefits:

Trees Saved ▸ **Tons of 100% Recycled Paper (From F)** [ ] x **24** trees per ton = [ ] Trees

Landfill Space Saved ▸ [ ] x **3** cubic yards per ton = [ ] Cubic Yards

Air Pollution Avoided ▸ [ ] x **58** pounds per ton = [ ] Pounds

157 Appendix 5: Designer Impact Analysis Form

Index

Acid-free papers, 106
Acid-sized papers, 100, 105–107
Air pollution, 6, 66, 119
Alkaline papers, 100, 105–107
Alum-rosin sizing, 105
American National Standards Institute, 106
American Paper Institute
   commitment to expansion of recycling, 17
   definition of mill broke (1980), 70
   projection of wastepaper utilization rate, 16–17
   use of recycling symbol, 74–75
   wastepaper recovery rate, 42, 126
American Society of Testing and Materials, 79–80, 106
American Soybean Association, 118–119
Anderson, Gary, 75
Appleton Papers, 55–56
Aqueous coatings, 115
Archival papers. See Permanence
Ash content, 46
Avoided costs, 65

Bagasse, 30
Basis weights, 124, 132
Bindings, 115–116
Bleaching
   of cotton pulp, 62
   of deinked stock, 53
   effect on brightness, 104–105
   with hydrogen peroxide, 105, 107
   with sodium hypochlorite, 67, 107
   of virgin and recycled pulps, 66–67
Brightness, 103–105, 132
Bulk, 94–95

Calcium carbonate, 46, 93, 106
Calendering, 95, 132
California, 76–77, 85
Canada, 122–123. See also EcoLogo Environmental Choice Program, 90–91
   paper mills in, 55–56
Canadian Standards Association, 90–91
Cellulose content
   in cotton and linen, 26, 62
   in wood, 31–32
Cellulose fiber, 46, 48, 58, 64, 94, 106
Chemical pulps, 132. See also Sulfate pulp, Sulfite pulp
   bleaching of, 66–67
   methods for producing, 32–33, 137
   and permanence, 105–107
   for printing and writing papers, 18
   recyclability of papers from, 113
   tonnage of trees to produce, 64
China, 21–22, 25, 122–123
Chlorine, 66–67, 105. See also Bleaching
Claudia Thompson Design, 112
Clean Air Act (1990), 119
Clear-cutting, 10–11, 14
Coated paper, 46–47, 97, 99–100, 119
Communications technologies and computers, 10, 12–13
Conservatree, 85, 88–89
Containerboard. See Paperboard
Container Corporation of America, 75
Contaminants
   affecting chemical bonds, 97
   from graphic design products, 113–117
   from paper additives, 105
   in wastepaper, 49, 53, 61

Converting wastes
   from independent dealers, 38, 45
   from paper mills, 3, 38, 45
   for recycled content, 70–73
Copiers. See Photocopying
Corrugated containers, 1, 8–9, 40–41. See also Old corrugated containers, Paperboard
Cotton fiber. See also Rags
   current use of, 62–63
   historic use of, 21, 25–28, 31
   and permanence, 105–106
Cotton linters, 63
Crane Paper Company, 31, 63
Cross Pointe Paper Corporation, 36–37, 52, 54–55, 60
C-stain, 107
Currency papers, 31, 62

Dandy roll, 101, 132
Deforestation, 10–14.
See also Wood consumption
Deinked pulp
   compared to virgin, 64–67
   effect on brightness, 104
   as market pulp, 56–57
   from 100% wastepaper, 36–37
   as postconsumer content, 82–84
Deinking, 132
   capacity expansions, 17–18
   invention and evolution of, 28, 33–35
Deinking plants
   cost of construction, 18, 49, 79
   cost of energy in, 65–66
   recyclability of wastes for, 113–117
   tax incentives for, 65
   in United States and Canada, 54–57
   use of office wastes for, 60

Deinking process, 48–53, 58, 137
Deinking sludge, 58–59, 65
Department of Commerce, 33–34
Desencrage Cascades Mill, 56
Designers. *See* Graphic designers
Didot, François, 28
Die-cutting, 117
Dioxins, 133
   from bleaching of pulps, 66–67, 105
   from incineration, 6
Direct-entry deinking, 58
Direct mail advertising, 10, 13
Dirt count, 60, 62
Dispersion, 49–50, 53
Domtar Papers, 61
Dot gain, 100–101, 133

Earth Day, 13, 86
Eastern Fine Paper Company, 73
EcoLogo, 47, 85, 90–91
Egypt, 21–22, 24, 31
Embossing, 117
Employment generated by recycling, 67
Energy consumption
   in dispersion process, 53
   in production of wood pulp, 38
   in recycled vs. virgin pulp, 65–66
Environmental Defense Fund, 87
Environmental issues, 3–14
Environmental Protection Agency, 5, 42, 66, 80, 88
EPA Guideline
   compared to RAC proposals, 81
   explanation of, 69–73
   and paper mill waste, 45
   recycled content diagrammed, 85
Europe, 24–26, 122–123
Exports of wastepaper, 17, 42, 44

Facsimile machines, 10, 12–13
Federal Trade Commission, 84–85
Felt side, 101
Fiber-added papers, 61–62, 133
Fibrillation, 94, 133
Fillers, 46–48, 95, 97, 101, 104
Film coatings, 101
Flax pulp, 62. *See also* Linen fiber
Flotation, 49–52, 54–55
Fluorescent dyes, 46, 104–105
Foil stamping, 116
Formation, 95–97, 100–101, 133
Fourdrinier machines, 28–30
Free-sheet papers, 64, 106, 133
French Paper Company, 45, 55
FSC Paper Company, 55
Furnish, 48, 54, 133

Georgia Pacific Corporation, 54–55
Glassine paper, 115
Glues, 115–116
Government purchasing, 110
   EPA Guideline, 69–73
   percentage of paper consumed by, 19
   state laws and regulations, 75–78
Grain of paper, 102
Graphic designers, 98
   and paper consumption, 109–119
   products in MSW, 8–9
   role in waste management, 124–127
Green Cross, 87–88
Green Report II, The, 84–86
Green Seal, 86–87
Groundwood pulp and paper, 32–33, 64, 106–107, 134, 137
Gutenberg, Johann, 24

Halftones, 100–101, 104, 114, 119
Handmade paper, 12, 21–28, 30, 46
Hard white envelope cuttings, 38
Hayes, Denis, 86
Hazardous wastes in MSW, 5–6
Heavy metals
   in deinking sludges, 59
   in incinerator ash, 6
   in leachates, 5
   in printing inks, 114, 119
Hemp, 30
Herbicides and clear-cutting, 10–11
Hickeys, 98–100
High-grade deinking wastes, 40
   in printing and writing papers, 41, 82
   in types of paper, 39, 41, 43–44
Hollander beater, 25

Incinerators, 4–6, 58–59, 109, 114
Ink coverage, 95, 99–101
   by design, 113–114
Inks, 114, 118–119
Institute for Scrap Recycling Industries, 38
Integrated paper mills, 45, 67
International Standards Organization, 106

Japan, 22–23, 122–123
Jasperson, Thomas, 33

Koops, Matthias, 27–28
Kraft paper, 33, 40, 66
Kraft pulp. *See* Sulfate pulp

Laid finish, 25, 101
Landfarming, 59
Landfills, 4–6
    for deinking sludges, 59
    heavy metals in, 114
    wastepaper in, 58–59, 109
Laser print, 40, 49, 59–60, 124. See also Office waste, Photocopying
Lignin, 31–32, 134
    effect on permanence, 106–107
    in groundwood papers, 64
    not in cotton, 62
Lincoln Pulp and Paper Company, 73
Linen fiber, 21, 25–28, 31, 105
Linting, 98–100, 103
Lloyd, William, 75

Makeready, 101, 103, 112–113, 134
Manila paper, 30
Manistique Papers, 55
Marcin, Anthony, 75
Market pulp, 56–57, 68, 136
Mechanical pulp. See Groundwood pulp and paper
Miami Paper Company, 45, 54–55, 60
Mill broke, 70, 77–78, 81, 84, 128
Mississippi River Corporation, 57
Mixed wastes, 40
Moist oven aging test, 107
Mould, 25–26, 28
Municipal solid waste (MSW), 4–9, 65, 134
    collection of paper from, 42, 126
    and deinking capacity, 17–18
    paper in, 6–9, 60

National Association of Printing Ink Manufacturers, 118
National Association of State Purchasing Officials, 79–80
National Information Standards Organization, 106
National Recycling Coalition, 80–81
National standard for recycled content
    argument for a, 128–129
    development of, 78–81
    and marketing claims, 84–85

Newsprint, 1. See also Old newsprint
    consumption of, 7–8, 12–13
    EPA Guideline for, 71
    groundwood pulp used for, 32
    in MSW, 7–9
    from ONP, 40–42
    wastepaper utilization rate for, 16–17
New York State, 77–78, 85, 91
Non-deinked wastes, 61
Noranda Forest Recycled Papers, 55–56
North America, 122–123

Office waste, 37, 40, 59–60, 109
Offset printing, 1, 97–103, 109
Ohio Pulp Mills, 56
Old corrugated containers (OCC), 7, 9, 39–41, 134
Old magazines (OMG), 41
Old newsprint (ONP), 7, 9, 40–41, 109, 125, 134
Opacity, 94–96, 104, 134

Packaging, 7–9, 12, 125
Packaging papers, 1, 12–13, 71
Paper bag consumption, 124
Paperboard, 1
    consumption of, 12–13
    EPA Guideline for, 71
    in MSW, 9
    use of wastepaper in, 42–44
    wastepaper utilization rate for, 16–17
Paper claims, 98–99, 102–103
Paper consumption, 4, 8–10, 12–13, 19
    designers' impact, 110–113
    equation for determining, 112–113
    and fiber shortages, 26, 30–31
    U.S. and foreign compared, 122
    and waste management hierarchy, 124–127
Paper machines
    invention of, 28–30
    speed of, 29, 33, 116
Paper mill trimming and converting wastes, 44–45, 70–73, 82
Paper strength, 96, 102
Papyrus, 21–22
Parchment, 22, 24
Patagonia Corporation, 110–111
Patriot Paper Corporation, 55–56, 60
Perfect binding, 115, 134
Permanence, 105–107

Pesticides, 10–11
P.H. Glatfelter Company, 45, 54–55
Phloroglucinol, 107
Photocopying, 124. See also Laser print, Office waste
    effects on deinking, 49, 59–60
    and paper consumption, 10, 12–13
Picking, 98–100, 103
Pick test, 99
Pigments, 114
Planchettes, 61–62
Plant materials as fiber sources, 22, 26–28, 30–33, 62–63
Plastic laminates, 115
Ponderosa Fibres of America, 56–57, 82–83, 115
Postconsumer waste, 85, 97, 126, 135
    in California regulations, 76–77
    in Conservatree definition, 89
    debate about, 81–84
    in EcoLogo program, 90
    in EPA Guideline, 70–73
    in Green Cross definition, 88
    in New York State regulations, 78
    in Recycling Advisory Council deliberations, 81
    in *The Green Report II*, 84–85
Postindustrial waste, 73
Postmill waste, 73, 85, 89, 112, 126, 128, 135
Preconsumer waste, 82–83, 134
    in API recovery rate, 42
    in EPA Guideline, 70–73
    illustrated in chart, 85
    in *The Green Report II*, 84–85
Pressure-sensitive adhesives, 116
Printing, 22, 24, 26
    problems associated with, 97–103
Printing and writing papers, 1, 8
    consumption of, 3, 8–10, 12–13
    composition of, 46
    deinked pulp for, 54–58
    EPA Guideline for, 71
    in MSW, 7–9
    in office wastepaper, 59–61
    recycled content standards for, 69–91
    wastepaper utilization for, 16–17, 42–45
Private certifications programs, 85–89
Private sector paper purchasing, 19, 110–112
Pulp substitutes, 38–39, 41, 43–44, 82
Pulpwood consumption, 14, 64

Rags, 25–28, 30–34
Réaumur, René Antoine Ferchault de, 27
Recovery rates for paper, 17, 42, 126–127
Recyclability, 77, 113–117
Recycled content
    in California regulations, 76–77
    cotton in, 63
    debate about, 78–85
    in EcoLogo program, 90–91
    in EPA Guideline, 69–73
    minimum standards for, 128
    percentages, 37, 46–48, 71
    in printing and writing papers, 16–17, 43–45
    symbol no guarantee of, 74–75
Recycling Advisory Council, 80–81
Recycling in waste management, 126–127
Recycling symbol, 74–75
Redfield, William C., 12, 33–34
Refining of fiber, 94–97, 135
Registration, in offset printing, 102
Resource Conservation and Recovery Act (1976), 69–70, 73, 135
Resource recovery plants. *See* Incinerators
Reuse in waste management, 125
Rittenhouse, William, 26
Riverside Paper Company, 55
Robert, Nicholas-Louis, 28–29

Sawdust, 73, 77, 84–85, 89
Scientific Certification Systems, 87
Secondary materials, 78, 135
Secondary waste, 73, 76–77, 135
Semi-chemical pulp processes, 137
Silk, 22
Simpson Paper Company, 55, 61
Sizing, 135
    alkaline or neutral, 105–107
    alum-rosin, 105
    purposes of, 26, 46
    surface, 98, 101
Sludge. *See* Deinking sludge
Soda pulp, 32
Solid waste. *See* Municipal Solid Waste
Source reduction, 5, 124, 135
Soviet Union, 122
Soybean inks, 118–119

State laws and regulations, 75–78
    in California, 76–77
    in New York, 77–78
Straw, 26, 28, 30
Stickies, 53, 115–116. *See also* Contaminants
Stick-on notes, 116
Sulfate pulp, 12, 32–33, 67, 137
Sulfite pulp, 12, 32–33, 137
Superfund National Priority List, 5
Swatchbooks, 112, 125

Tape pull test, 99
Tax incentives
    encouraging recycling, 76
    supporting the timber industry, 64–65
Thermography, 117
Tipping fees, 4, 6, 65
Tissue papers, 1. *See also* Toilet paper
    consumption of, 8, 12–13
    deinked pulp for, 57
    EPA Guideline for, 71
    in MSW, 9
    use of wastepaper in, 39, 41–44
    wastepaper utilization rate for, 16–17
Titanium dioxide, 46, 105
Toilet paper, 12, 41, 71, 126
Tree farming, 10, 13–14
Ts'ai Lun, 21–22

Uncoated paper, 46–47, 54–55, 97, 99–101, 119
Upcycling, 41
U.S. Forest Service, 11, 64–65
UV coatings, 114–115

Varnishes, 117
Vegetable inks, 118–119
Vellum, 22, 24
Virgin wood pulp
    compared with recycled pulp, 64–67, 94–95, 97, 104, 107
    use of, 15, 18, 34–35, 43
Volatile organic compounds, 119

Washing, 49–52, 54
Wasps, 27–28
Waste management hierarchy, 124–127
Wastepaper, 37–40, 42–44, 49, 114

Wastepaper collection, 17, 42, 60, 113, 126
Wastepaper utilization rates, 14–15, 17, 127, 135
    in newsprint, 16–17
    in paperboard, 16–17
    in printing and writing papers, 16–17, 43–44
    U.S. and foreign compared, 122–123
    during World War II, 34–35, 127
Waste-to-energy plants. *See* Incinerators
Web papers, 102
WGBH Educational Foundation, 111
Wire side, 101
Wool, 25–26
Wood consumption, 10–11, 14
    from public lands, 11, 64–65
    trees per ton of paper, 64
Wood-free pulps, 106, 133
Wood shives, 61
World wars, 33–35

Colophon

Paper
To complete this project we consumed 537 pounds of office paper for copying information, printing and distributing drafts, correspondence, book design, and keeping office records. This total does not include the amount of paper generated on our behalf from other sources, such as information, correspondence, subscriptions, publications, and paper samples sent or otherwise provided to us.

Our first printing of 12,500 copies required 14.3 tons of paper, including 412 pounds of paper used for press proofs to evaluate ink choices and halftone reproduction. The following papers were used:

Cover (Paperback edition)
Simpson Quest Cover Plus, 88 lb., Putty. The fiber content for this paper is 100% recycled. It is obtained from printed wastepaper that is used directly without deinking; this results in the flecks that give the paper its textured appearance. It is an acid-free paper made to a neutral pH.

Frontmatter to Page 2
Mohawk P/C Text, 70 lb., Colonial White, Vellum finish. The fiber content is a minimum of 60% recycled. The majority of this recycled content comes from purchased pulp substitutes; one-sixth of the recycled fiber, or 10% of the paper's total fiber content, comes from lightly printed wastepaper that is not deinked but is used directly. The paper has a minimum pH of 7.5 and minimum 2% alkaline reserve.

Pages 3 to 130
Cross Pointe Troy Book, 80 lb., Blue White, Satin finish. A special making order of Troy Book was produced for this publication from 100% deinked recycled fiber, all of which was derived from printed wastepaper processed through Cross Pointe's deinking plant. This paper has a minimum pH of 7.5 and a minimum 2% alkaline reserve.

Pages 131 to 162
French Speckletone Text, 70 lb., Madero Beach. This paper contains a minimum of 75% recycled fiber, the majority of which comes from purchased pulp substitutes. At least 15% of the paper's total fiber content was derived from deinked wastepaper, through the purchase of deinked market pulp. The paper has a minimum pH of 7.5 and a minimum 2% alkaline reserve. This is a fiber-added paper; wood shives give it its textured appearance.

Printing
The book was printed by The Stinehour Press in Lunenberg, Vermont. The photographic images were reproduced from both black and white photographic prints and from color transparencies. Black and white prints were screened using conventional camera techniques with a 300-line screen. The transparencies were scanned on a Dainippon DS 608 laser scanner at 200 lines per inch. The text pages were printed on a Heidelberg SORS/Z 28x40" two-color offset press. Paperback covers were printed on a Heidelberg Speedmaster 28x40" four-color press.

The inks used would typically be marketed as "soy-based" inks. The blue ink used throughout has an oil content that is over 75% soybean oil, with the remaining oil content petroleum-based. The oil content of the black ink is just over 50% soybean oil. The inks used on the cover of the paperback edition have an oil content that ranges from 50 to 80% soy-based.

Design and Production
The book was designed by Claudia Thompson and Susan Larocque and produced on an Apple Macintosh computer using Aldus Pagemaker 4.01 and Adobe Illustrator 3.0.1. Page mechanicals were output on a Varityper 5300 at 2400 dots per inch. Overlays for screens and color were cut by hand as necessary.

The fonts used, Zurich and Classical Garamond, were drawn for the Macintosh by Bitstream Corporation. Zurich is based on Adrian Frutiger's design of the sans-serif Univers type family in the 1950s. Classical Garamond is based on Jan Tschichold's typeface Sabon, designed in the 1960s and inspired by typefaces originally cut by Claude Garamond in the mid-1500s.